D1623936

Human Frontiers

Also by Michael Bhaskar

The Content Machine
Curation
The Oxford Handbook of Publishing

Michael Bhaskar

Human Frontiers

The Future of Big Ideas in an Age of Small Thinking

The MIT Press
Cambridge, Massachusetts
London, England

First published in the United Kingdom by The Bridge Street Press,
an imprint of Little, Brown Book Group Limited.

This book was set in Sabon by M Rules.
Printed and bound in the United States of America.

Library of Congress Cataloging-in-Publication Data is available.

ISBN: 978-0-262-04638-1

10 9 8 7 6 5 4 3 2 1

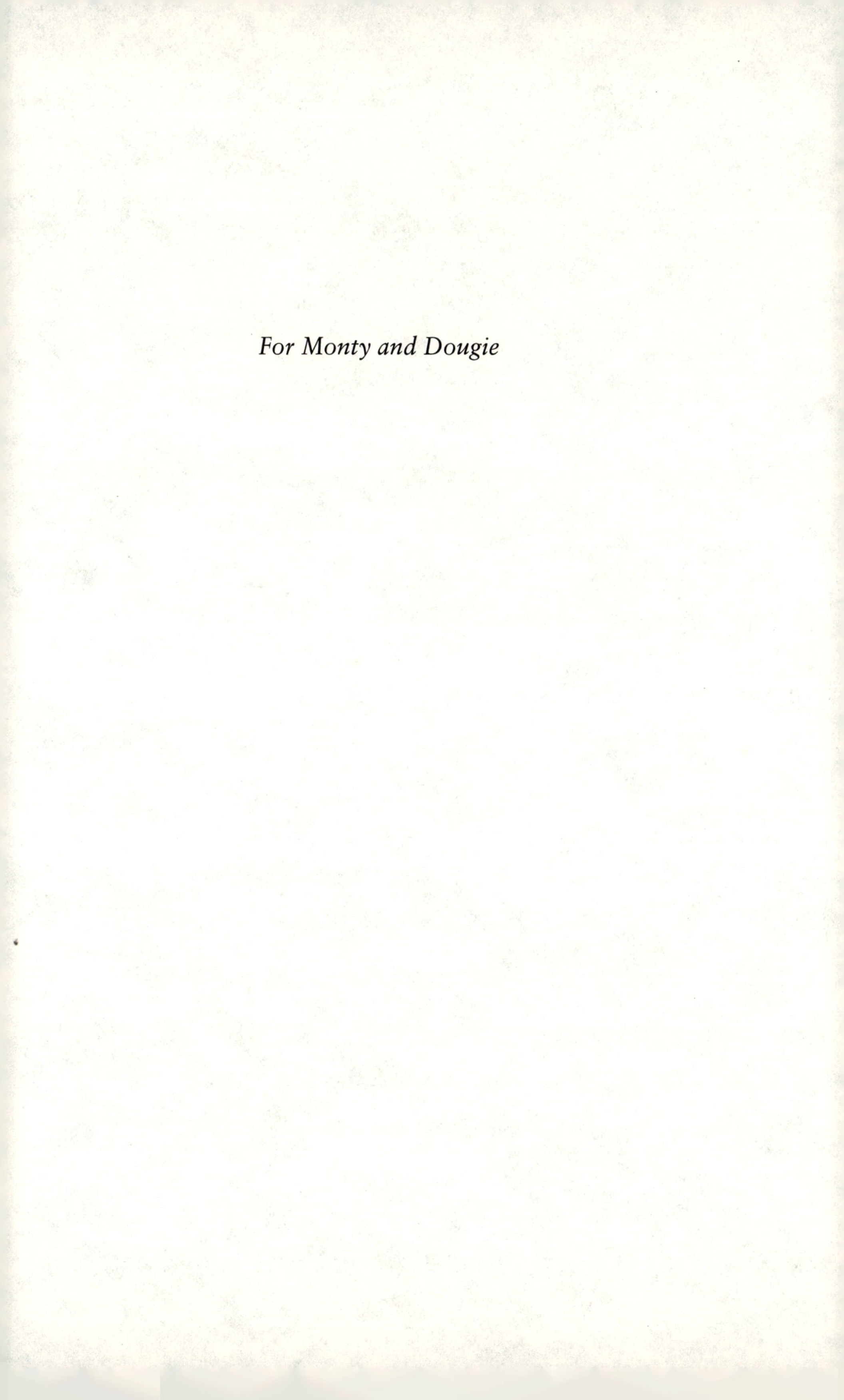

For Monty and Dougie

Contents

‘As the investigation goes on, additions to our knowledge cost more and more, and, at the same time, are of less and less worth.’

C.S. PEIRCE

‘I fear that the mind may keep folding itself up in a narrower compass forever without producing new ideas, that men will wear themselves out in trivial, lonely, futile activity, and that for all its constant agitation humanity will make no advance.’

ALEXIS DE TOCQUEVILLE

‘Ideas are powerful. Ideas drive change.’

ABHIJIT V. BANNERJEE
AND ESTHER DUFLO

Prologue

Archimedes was worried. Hieron II, King of Syracuse, had summoned him several weeks before. Always with that slightly baleful look in his eye. Straight away, Archimedes knew it was going to be difficult. Keep your head down, he told himself. Stay out of trouble. In fact, why not go back to Alexandria and work in peace at the Library?

It turned out Hieron had been given a golden crown. Wonderful, said Archimedes, most justly deserved!

But there was a problem. Hieron was convinced he'd been swindled, that the golden crown was adulterated with silver. He needed, absolutely needed to know for sure. OK, thought Archimedes. If we melt the crown we can check its volume, which will tell us whether or not it's pure gold.

Melt the crown! Hieron almost hit the roof. Melt the crown? Laughable, disgusting, impossible. Not going to happen. Go and figure it out, said Hieron. You're clever.

So that was what Archimedes had tried. But all the usual tricks didn't work. He couldn't concentrate on his other projects: a paper on geometry, refinements of a design for a new kind of pulley. He paced the sea-lapped city walls, felt the breeze and brooded on the implied threat in Hieron's words; the implied threat despite all he'd done for Syracuse, and on a matter so pointless. He sighed: this was the problem with tyrants.

Archimedes had always been fastidiously clean and in the chaos of papyrus, parchment and models that was his household, the one moment of order was his daily bath. Time at last to forget about Hieron and his ridiculous requests.

That day, as Archimedes stepped into the bath, he noticed something both utterly trivial and yet extraordinary: as he got in, the water level rose. His body displaced the water. The volume of water displaced equalled the volume of his body.

He instantly knew two things: first, that he'd solved the riddle of the crown. Second, that the insight was of much greater importance: here was a universal means of gauging volume, a breakthrough moment.

'Eureka!' he cried, leaping from the bath. Eureka – I found it! The residents of Syracuse were used to the eccentric ways of the sage, but even for them the sight of this wizened man running naked through the central market and the port, past the temples and battlements, again and again shouting 'Eureka!', was unusual.

But Archimedes didn't care, this was no ordinary day – and no ordinary idea.

In hindsight it seems so simple. But like every good idea, it wasn't.

Everyone knows the story. When we think of big, breakthrough ideas, we think of something like this, the original, endlessly rehearsed 'Eureka' moment. A flash of insight, a crystalline idea, a new plane of thought. In one sense this is right, in that it suggests ideas aren't equal, that some ideas are more difficult, more significant, than others. In another sense it couldn't be further from the actual messy, grounded, discursive, agglutinative, coincidental, resource-hungry process of ideation.

It won't come as a surprise to hear that the above story is probably apocryphal. But in fact the truth of Archimedes and Eureka is far more instructive for the history of ideas.

Archimedes lived an extraordinary life, his interests ranging across the span of classical knowledge. His breakthroughs were often abstract: he helped invent the concept of pi and pioneered the mathematics of parabolas, levers, spheres, cones and irregular solids. We can't be sure of the bath and the crown; we can be sure that Archimedes was a world-leading expert in the study of fluids and mathematics.

But he was also intensely practical, not only an heir to Euclid but an engineer working at the forefront of Hellenistic technology. He built engines of war and mechanical inventions including pulleys, winches, the gear wheel and the hydraulic screw. He combined high-level academic work with a hands-on approach, evident in the unification of astronomical discovery with mechanical genius in the creation of one of the world's first orreries. Each alone would make him a giant of antiquity.

In the third century BCE, Syracuse, in Sicily, was the capital of Magna Graecia – greater Greece – and of the entire central Mediterranean. It lay at the centre of Greek culture and trade, a place where Aeschylus wrote plays, a city frequented by Plato. Embroiled in the epic Second Punic War between Carthage and Rome, Syracuse sided with the Carthaginians. It was a mistake. But as the might of Rome was thrown against the city, Archimedes the astronomer and scholar threw himself into its defence, creating a system of powerful countermeasures.

Such defences had never been seen before: giant mirrors to deflect the sun's rays onto enemy ships, supposedly setting them alight; powerful torsion siege engines; large, castellated warships; a kind of vast wooden claw, balanced with counterweights, capable of lifting and overturning enemy ships as they approached the city's sea walls. In Plutarch's words, the Romans saw him as a 'magician', capable almost single-handed of staving off their attacks.[1] When, in 212 BCE, the Roman consul Marcellus stole into the city thanks to a treasonous Syracusan, Roman historians report that he gave two orders: that the royal treasure be respected and that Archimedes' life be spared. But, supposedly lost in a complex problem, the inventor was killed by a legionary ignorant of his identity.

Archimedes worked across a variety of fields, over many years, tackling problem after problem. His whole life was geared around technical and mathematical thinking. Archimedes had no doubt been circulating the crown problem in the back of his mind, a conscious and subconscious 'priming' that meant ideas were productively brought together. He had an extraordinary capacity for observation underpinned by extreme mental agility, coupling abstract mathematical concepts and synthesising them with a practical, insightful worldview. But recent scholarship makes clear that his was also 'a mind constantly engaged in perfecting and innovating pre-existing technology, rather than the isolated expression of a genius' activity'.[2] The Eureka moment was actually a slow, no doubt difficult process of cross-pollination as much as it was a sudden flash; the culmination of decades, not seconds.

We also know that Archimedes worked far from alone. He was educated by and deeply plugged into a wider research network. The pivotal role of Syracuse in the Mediterranean was helpful – it connected Archimedes to the Library of Alexandria, the great scholarly hub of the classical world. Archimedes studied under Conon of Samos in Alexandria and was familiar with thinkers of his age like Philo of Byzantium, the astronomer Aristarchus, who expounded a heliocentric cosmology, and the proto-scientist Eratosthenes, who calculated the circumference of the Earth.

We can take two things from Archimedes and the crown. First, that some ideas really do have outsized impact. They change the world. Whenever it was that humanity first robustly realised it could measure volume, it had entered a new mathematical, mental, technical realm. Our sphere of understanding and operation had been enlarged.

Second, that such breakthroughs do not, in fact, spring *ex nihilo* in the bath. Instead they are formed of previous ideas, remixed and morphed; they are built on networks of people, working sometimes together, sometimes at cross-purposes. They arise from both the realm of abstract thought and the expensive, pragmatic, tiring work of building city walls and operating siege engines. Big ideas have long and winding gestations, embedded in all the currents, networks and contradictions of their times. They are contextual, social and, because of that, highly contingent.

The real Archimedes is so much more instructive than the myth. If ideas were all abstracted Eureka moments, understanding them would at best become a kind of psychology. Instead ideas are amenable to wide-ranging explanation and study. We can follow their gestations, trajectories and interlinkages; map the conditions of their arrival and the flows of their flourishing and demise. Their very groundedness, whether technological, cultural, scientific, political or mercantile, means we can trace their development.

This lets us ask interesting questions: what ideas underpin our civilisation, and where are they heading? Are we living in a wonderland of innovation, or might we instead be experiencing a mirage of progress? Is this, around us, a society of big ideas or small horizons?

Archimedes and countless others took up the challenge of finding world-changing ideas. That search shaped history. It will shape humanity's next steps as well as our ultimate limits. It's time for a closer look.

Introduction

Life at the Human Frontier

People like Archimedes work at the limits of knowledge, at the frontier; the far boundary of what the human race knows at any one time. This isn't a hard border. Rather, like the original geographical frontiers, it is a shifting, debatable zone of what may be true or known, thought or possible.

Over the last millennium, and especially in the last couple of centuries, the frontier has been decisively pushed back – almost every year portions of it fell away to reveal new vistas. Humanity as a whole simply knows more now than at any other time. We have grown used to the knowledge frontier's steady recession; a bold history of exploration and achievement has felt like a natural part of the human order.

But this concept of a human frontier encompasses much more than knowledge alone. There are frontiers that mark how fast we can travel, how long we can live, what imaginative universes we can inhabit or produce for others; frontiers in the scale or wealth of our societies and enterprises. These many human frontiers define us. They too exist in an entrenched pattern of expansion. To understand humanity at the highest level is, at any given moment, to understand its frontiers, and further, to understand what is happening at and to those frontiers.[1]

What keeps them moving is ideas. They are the force nudging into the unknown, widening the domain of the possible. Most ideas achieve little, but some, like those of Archimedes, click with non-linear impact, shifting boundaries, opening territory. Their legacy is an unambiguous movement at the frontier. Discoveries in science; inventions in technology; creative change in the arts; pioneers in business or exploration – all of these are varieties of big ideas, that is, the ideas that have made the most impact at the frontier.

Boiled down, the history of humanity is the history of big ideas pushing those frontiers, from the wheel to space flight, cave painting to the massively multiplayer game, monotheistic religion to special relativity and universal suffrage. And the same is true of our future. What happens at the human frontier shapes human existence itself.

So how's it doing?

By the 1960s the developed world had witnessed two hundred years of extraordinary activity on every stretch of the frontier. From transcontinental flight to the defeat of once epic killers like tuberculosis; the invention of television to modernist art; the creation of cars, electric light, quantum mechanics, the welfare state, civil rights, genetics, nuclear power, jazz, the Beatles, organ transplants and the research university: no wonder the future had a utopian tinge. This was, after all, the era of the moonshot; of political obsession with technological mastery; of experimentation in music, culture, knowledge and relationships. Humanity had undergone a big ideas revolution and the expectation was that progress would keep accelerating.

In 1967 two American strategists, working at the forefront

of new thinking, wrote a book called *The Year 2000*.[2] They laid out a systematic overview of what would happen over the next thirty-three years. The project forecast forthcoming milestones in science and technology, innovations we might rationally expect by the year 2000, even producing a table of 100 technologies they assumed would be implemented.

Most remain unrealised. We do not have 'control of the weather and/or climate', a permanent manned lunar installation or even 'automated home maintenance'. There is very little 'extensive use of cyborg techniques' or 'undersea colonies'. We do not have a reliable way of reversing ageing, 'physically non harmful methods of overindulging' or 'individual flying platforms'. We do not directly communicate into the brain. Likewise, it is perhaps for the best that we don't have 'space defense platforms' or use nuclear weapons for mining.

Of their suggestions, we've got near to some, but in practice many are either too difficult or too mundane to care about. Synthetic foods? Well, there's soylent for those that want it. 'Controlled and/or super-effective relaxation and sleep'? There are drugs for those, but no one would say we have it nailed.

Sure, we have video cassette players and home computers (remember those?), satellite TV, 'extensive use of transplantation of human organs' and 'commercial extraction of oil from shale'. But it's hard not to feel that the authors, Herman Kahn and Anthony J. Wiener, anticipated something more radical and decisive. Years on from the millennium many, if not most, of their 'very likely technical innovations' have not been made despite the authors' belief they would in '90–95 per cent' of cases. What happened? And what does that tell us about the future?

*

Go back to the lives of your great-great-great-grandparents. They were probably alive towards the middle of the nineteenth century. Like you, they had hopes, fears and dreams. They craved what we crave: security, love, entertainment. Yet their world, just a generation or two out of reach, was more dangerous, backbreaking, boring, precarious and limited than anything we in the West ever experience, coronavirus and its aftermath included.

In the mid-nineteenth century, buildings were often wooden and rudimentary. In cities especially, this meant devastating fires were common. The boomtown of Chicago was almost eradicated in a vast urban conflagration in 1871. Sewage was a ubiquitous disease-carrying presence, in the home and out. Animals were everywhere, their dung clogging the streets. Those animals were scrawny and ill fed: a cow produced 1000 lb of milk a year, compared to 16,000 lb today.[3] Clean water and unspoiled food, monotonous and meagre though it might be, were rarities. Disease was rife and little understood. Between a fifth and a quarter of all infants died before their fifth birthday. Unimaginable personal tragedy was an unexplained but near inevitable part of life.

Most people never moved far from where they were born, and never encountered much outside their own small cultures. The railway had started to change that, but for the most part travel was, as it had been for millennia, dictated by the pace of hoof and sail. Likewise, although there were a few industrial jobs (in grim, dark, limb-mauling places), most people worked the land, using technology and methods that again changed at glacial pace. Illiteracy was the norm, access to knowledge and entertainment media scant. In the evenings people relied on the faint, expensive and polluting light of bad candles or whale oil. Most went to bed when it got dark.

Our recent ancestors, in some ways so close and so similar,

inhabited a very different world. But it had already begun changing at a blistering pace. In a little more than a generation, their sons and daughters and grandchildren would be living in a landscape that is recognisably modern.

For the first time in history there was exponential population growth: the US started the nineteenth century with just 5.3 million citizens but finished with 76 million, bigger than any European country save Russia. In the space of a few decades homes were transformed from quasi-medieval hovels into the 'networked' house: clean hot and cold water ran in and out. Electricity powered a host of 'electric servants' that began to take over some of the back-breaking domestic labour which had dominated the lives of women.

The lightbulb and the motor car were both effectively invented in 1879. At the end of the nineteenth century they were still novelties; twenty years later, both were produced by the million. And it wasn't just cars and lights: telephones, aeroplanes, canned and processed food, the modern corporation and production methods, radio, refrigeration and the first plastics all burst into society around this time. No wonder there was an unprecedented revolution in productivity. Inventions like Cyrus McCormick's threshing machine led to a 500 per cent increase in output of wheat per hour.[4] Isaac Singer's sewing machine meant the time spent making a shirt was reduced from over fourteen hours to just one hour and sixteen minutes.

Big ideas were conceived and executed at what felt like an accelerating pace. The bounds of knowledge shifted – fundamental forces like energy and evolution became tractable. The mystery of disease started to unravel. Just after the turn of the twentieth century, Albert Einstein transformed basic categories like time and space. Culture underwent a revolution. Mass entertainment, from television to radio shows, became a reality, while the nature of art was progressively redefined, from Impressionism to Abstract Expressionism. By the early twentieth century, people could believe neither their eyes nor their ears. The franchise was expanded; the first glimmers of a welfare state made themselves felt. Knowledge, culture, technology, social organisation, everyday life: each entered a revolutionary cycle.

Our great-great-great-grandparents experienced that rare thing: for perhaps the first time in history *every dimension* of their human frontier changed. This was a historic break, built on rapid, vaunting advances, decade after decade.

When Kahn and Wiener built their model, it was a natural continuation of this outpouring. It seemed inevitable that the frontiers would continue to gloriously unfurl.

But it's become clear that the trajectory is more complicated. Change is rapid in some areas, yet has slowed in others. Above the roiling everyday tempest of life, above even pandemics, crashes and the fissures and rivalries of nations, the future is once again uncertain. The frontier is, against expectations, getting stuck.

Welcome to the great meta-problem of the twenty-first century.

We're at a curious juncture of history. To open the latest *Wired*, *Nature* or cultural review is to be treated to a parade of wonders: extraordinary new discoveries, products and achievements. From quantum biology, nanotechnology and exoplanet astronomy to nudge unit governance, blockchain and virtual worlds, surely this is a uniquely fecund moment when the frontier is expanding at a record and still accelerating pace.

Mounting evidence suggests the opposite. A lively debate is calling into question these dominant assumptions about our place in history and the default nature of progress. As much as anyone, the economist Tyler Cowen rang the alarm when he called the present, particularly in the West, the Great Stagnation. So let's call it the Great Stagnation Debate.

Nicholas Negroponte believes we exist amid a 'Big Idea Famine'. Others see not so much an innovation machine as an 'Innovation Illusion'. The economist Robert Gordon talks about the 'fall of growth', while the physicist Lee Smolin talks about a science no longer capable of revolutionary thought. These and other thinkers point to flat growth rates; incremental and derivative technology; paradigms of knowledge and culture that have been stuck for decades. The common perception of acceleration, they argue, no longer applies to major revolutions. Rather, we move only in cautious increments. In response to great challenges, these critics see 'complacency' and 'decadence'. After fifty years we haven't been back to the Moon; we haven't cured cancer; life expectancy increases have stalled; we're still horrifically addicted to carbon-based energy. We didn't realise the visions of Kahn, Wiener and the postwar world.

Our societies have been floored by a widely predicted pandemic. As I worked on the book, Covid-19 overturned much we took for granted, from meals at restaurants to transatlantic flights to safe government debt levels. But in the face of perhaps the biggest challenge for seventy-five years, government, corporate and even personal thinking was often trapped by the models of the past, incapable of building those of the future on the fly. Despite all our technologies, businesses and knowledge, we are vulnerable.

Entrepreneur and investor Peter Thiel famously encapsulated the argument as 'We wanted flying cars, instead we got 140 characters.' But there is more to it than that – in the words of David Graeber, it encompasses 'a profound sense of disappointment about the nature of the world we live in, a sense of a broken promise – of a solemn promise we felt we were given as children about what the adult world was supposed to be like'.[5] This isn't just about flying cars and colonies on Mars, the tug of war between techno optimism and pessimism. My view here is broader than technology or economics, significant as they are; it's about a world that is, to paraphrase Ross Douthat, one of 'stagnation, sterility, sclerosis and repetition'.[6] Stagnationists come from across the intellectual spectrum: free market libertarian entrepreneurs, social conservative newspaper columnists, centrist business consultants and radical anarchist academics; artists, architects and anthropologists as well as economists and tech titans.

The debate suggests, despite our many advantages, that there is a slowing down at the frontier. But the thesis is far from universally accepted – there is plenty of evidence to the contrary, and in the face of a rush of discoveries even some of its progenitors are wondering if there is a reversal. After all, the pandemic has also been a time of super-fast vaccines, reimagined logistics, habits changed on a grand scale.

Before thinking of the future, then, we need to ask: are we really delivering fewer ideas capable of smashing through those human frontiers? How is the frontier stickier, slower, further away, more dispersed, and what does that mean? Why is it, when we have more people, with access to more knowledge and tools than ever, notionally spending their time on thinking, researching and creating, that there are signs of a slowdown in big ideas?

Above those, however, is perhaps the ultimate question of the Anthropocene: whether this trend of relative slowdown, of big ideas stumbling rather than soaring, will continue and intensify over the next century, or whether a series of efforts and initiatives can reverse it. I will look to the most likely scenarios here, exploring everything from the structure of knowledge to the regulation of innovations, the growth of education in China to the boom in AI research. Put simply, can we overcome the increased challenge, pick up the pace and keep rolling back the frontier, or are we fated to carry on as our ideas inexorably get tougher, our frontiers ossify and decay?

The future of big ideas matters. It matters if we are to beat the next pandemic. It matters if you get ill and, thanks to antibiotic resistance, nothing can make you better. It matters when a family member succumbs to dementia. When the seas rise and our cities flood, we'll pray for a big idea in energy or climate science. More than anything else, ideas define us: they are at the heart of our art, politics, culture, science, technology and economy. They lie behind the stories we tell, the buildings we inhabit, the goods we consume, the things we know or believe we know.

At this delicate juncture in history, we need huge new forms of collective invention, not least to meet the severe multigenerational challenge of climate change. We need a vision for a world poised to fulfil our potential. A world willing to think anew, beyond the known or imagined. A world that can paint on a daring, radical, different canvas. A world where the human frontier is alive, roiling, moving, expanding.

My journey to thinking about the human frontier has not been straightforward. My background is in the humanities, literature and publishing, which are not obvious starting points. But some years ago, watching technological change in the creative industries up close, I started to write about the history, theory and practice of technological, business and cultural transformation over a wide span of time and space, from ancient China to Silicon Valley. My research opened to a huge variety of interlocutors and perspectives. At the same time my work with writers had given me an enduring fascination with how people think and conjure the completely new.

I began to study the role of imagination in society, from scientists to entrepreneurs, before getting sidetracked by an opportunity to work on the scientific and technical frontier as a consultant writer in one of the world's foremost technological research organisations. Fascinating in its own right, it gave me a grounding at the limits of contemporary ideas. It also set me thinking about the role of specific organisations in catalysing imagination, invention and innovation. This research in turn led me to the Great Stagnation Debate, a lively if disjointed discussion in peer-reviewed economics journals and on blogs, in TED lectures and on Twitter. Although largely overlooked in mainstream discussions, it felt like the biggest story of our time. That debate resonated with me, addressing a nagging but vague feeling I'd had for years.

Yet it was equally clear that there were no simple answers – to start with, there's no established methodology or criterion for settling it. What really mattered about stagnation was our destination. That was surely worth exploring.

Because the subject matter was so diffuse, elements of the debate were also failing to connect with each other. Physicists might talk to technologists, but were less engaged with economists, and even less so with cultural or political theorists. Economics professors were unlikely to cite anecdotal evidence from chemistry lecturers. Some writers were attentive to social or business matters, but then left out large portions of technological research; others pored over technology but forgot the cultural landscape. I wanted to see them all as aspects of a broad, interrelated phenomenon. Bridging these scattered conversations would bring the debate – and the future – into focus. The more I researched, the more this great interlinkage became evident. Outside disciplinary bubbles, the big picture naturally started to emerge.

But how to approach a topic like this – both significant to every field of endeavour and, because of that, so vast as to be formless and unapproachable? Behind all this splintered activity and discussion, the common unit was the idea. Here was a way to understand progress, innovation, stagnation and slowdown. At the same time I kept coming across the concept of the knowledge frontier. Just as ideas were the common thread, so was this image of the frontier, and what's more they were linked; humanity's various frontiers depended on and moved because of ideas. The makings of a project were in place, one I saw as a sort of global history of ideas, shifted into the present, with one eye on the future.

By this time I'd moved some way from the starting points, both in years and scope. Frequently swimming out of my depth, it was an odd, daunting but exhilarating space for an independent writer, scholar and publisher. I was fortunate to rely on hundreds of excellent books and scholarly papers and scores of conversations and exchanges with some of the world's most interesting thinkers. I was surprised and gratified that everyone had an opinion on the subject, that it was one people instinctively wanted to discuss; we all have our own sensations of acceleration and stasis.

The book itself is divided into two parts. In Part I, I look at what is meant by a big idea, before reviewing the Great Stagnation Debate. I examine the evidence for a slowdown in big ideas, and explore how widespread it is. Then in Part II I move to a more diagnostic and forward-looking mode, explaining the causes while projecting them into the future.

There I focus on two major blockers to big ideas – respectively, problems inherent in the direction of ideas and problems with our society. First, the nature of ideas means we tackle easier ones first, so that over time we're left with harder, more intractable challenges. At the same time infinite development or discovery isn't always possible; eventually we reach the limits, subjective, economic or physical. In some areas there is a ceiling to new big ideas, and we are inescapably closer to that ceiling than ever before. Second, society has become more hostile to radical innovation, risk-averse, fractious, short-termist. Entrenched interests, a supercharged financialism, swollen bureaucracies, resurgent populisms: all inhibit the most daring forms of new thinking. Raised as we are on a diet of techno-boosterism, this often comes as a surprise. Nonetheless we have built a cautious and unimaginative world.

We then consider two major forces that could get things moving. For the first time in history, as the developing world closes the gap with advanced economies, all major civilisations are operating at the frontier of knowledge. Equally, we could be at the cusp of a revolution in tools and technology as profound as that of the late nineteenth and early twentieth centuries. AI, quantum computing, 3D printing, synthetic biology, neuralink interfaces connecting computers and human brains: these are not just buzzwords, but potentially transformatory. Those forces are then weighed in the balance, before I put forward suggestions about what would help push back the frontier and meet the challenge of our times. Overall I am cautiously and conditionally optimistic. Warning lights have been flashing throughout the early twenty-first century, but unprecedented technologies and civilisations of awesome scope might also await us; reason enough to feel that big ideas are not done yet.

Lastly, in the epilogue we move firmly into speculative territory, to consider what future changes might be wrought in our moral precepts, cosmology, religious beliefs, ability to harness energy, domains of study, forms of business and political economy.

Coming up with groundbreaking ideas has never been easy. On the contrary, they have always been rare bright spots in seas of ignorance and hardship. We aren't that different to Archimedes and his peers, facing the unknown. The most extreme forms of creativity, business, knowledge and achievement are still exceptionally difficult. But that doesn't mean we can't do better. As this book shows, we can and we should expect more.

This is the story of Archimedes brought into the present, and beyond. The story of whether our advance to the human frontier and its future, and everything that comes with it, is accelerating or slowing down.

Part I

BIG IDEAS TODAY

1

How Big Ideas Work

A Better Idea

For most of recorded history (about 97 per cent, in fact), not much changed over the course of the average lifetime.[1] The future of ideas was much like its past. Yet by the nineteenth century this was no longer true. Since the late eighteenth century, per capita GDP has risen by up to 10,000 per cent in the richest parts of the world, an astonishing and completely unprecedented change.[2] Explaining this vast increase – what has been called the Great Enrichment, or the Great Divergence, when a handful of north-west European economies hit the accelerator – is one of history's great questions.

Over the years historians and economists have offered a variety of answers. Proposed causes include new patterns of trade; the discovery and use of natural resources, principally coal; the role of imperialism; even simple economic accumulation. Perhaps it was a better class of institutions, encouraging positive attitudes to private property and entrepreneurialism. Urbanisation, agricultural enclosure, wars and international competition, demographic shifts – all have been touted.

And yet coal had sat in the ground for eons without igniting such a revolution. Global trade networks long predated the eighteenth century. Quantitative economic measurements are inadequate to explain growth in the order of one hundred-fold in the wealthiest nations. Behind those material changes lies something even more profound. In the words of the economic historian Deirdre Nansen McCloskey, 'Our riches did *not* come from piling brick on brick, or bachelor's degree on bachelor's degree, or bank balance on bank balance, but from piling idea on idea.'[3] The most important transition in recent history was built on ideas.

This profusion of new ideas rested on new attitudes towards ideas; in other words, on ideas about ideas.[4] For much of history the dominant assumption was that everything had already been thought, that ancient authorities like Confucius or Aristotle or Jesus had the final word. Originality was dangerous and probably impossible.

But around the sixteenth and seventeenth century, Europe changed, attitudes shifted. Now original thoughts were to be celebrated. An embrace of ideas ignited a process of further openness and innovation, widely diffused and ongoing, and unlike in, say, ancient Athens or Song Dynasty China it became a self-sustaining change. In Britain and the Netherlands especially, there was a sense that a wide range of people could now seriously propose new ideas and that they should be praised and valued for doing so.

This mentality was everywhere from architecture to the making of clothes. Anton Howes points out that the invention of John Kay's flying shuttle, an innovation that made cotton weaving far more efficient, could have happened at any time in thousands of years.[5] All it required was wood and string. It was a practical step that did not rest on specialist technical or scientific insight. Yet until 1733 no one had fully thought to do it. Many of the key technologies of the time were arguably similar, Richard Arkwright's water frame, one of the central industrial inventions, among them. Jethro Tull's seed drill had antecedents in ancient China and even Mesopotamia, but only found firm purchase in the English countryside of 1701.

For the first time, the publication of results was seen as a positive. Innovators shared their new mechanisms directly with their peers and with government organisations, in periodicals and at public displays. Membership of bodies like the Royal Society and the Royal Society of Arts, Manufactures and Commerce was desirable. Steadily this ecosystem gave rise to the 'first knowledge economy', one built on a culture of ideas creation. Not coincidentally, this was the first time an economy hit escape velocity.[6]

In the words of economic historian Joel Mokyr, this was a 'culture of growth'.[7] A new proto-scientific culture spanned Europe: a Republic of Letters, an 'invisible college' and transnational market for ideas, where leading thinkers created a buzzing epistolary network devoted to discovery, exploration, thought and experiment. Led by path-breaking cultural entrepreneurs – people like the statesman and natural philosopher Francis Bacon, who encouraged a new empirical, improving attitude – the point of this 'college' was to proffer something original and unknown. Its leading figures, like Isaac Newton, Gottfried Wilhelm Leibniz, John Locke and Voltaire, created an economy of prestige which acted as an incentive for new entrants to try their hand.

The culture of ideas, once closed, was now open. Curiosity wasn't weird or sinful. The natural world was subject to reason, manipulation; the domain of humanity, not its master. Out of this republic grew the seminal insight that knowledge about the world could both buttress and be informed by technology, and that, crucially, a dialogue between the two could create something shocking and elusive: progress, no longer a chimera or shunted up to the afterlife, but possible on Earth. Conditions were in place for the Great Enrichment. Got some coal? Now there was something to do with it. Want to trade? You had new, much more cost-efficient goods with which to do so.

Some changes were wrought by those without scientific backgrounds – think of George Stephenson and his *Rocket* – but science did still contribute to technology, as with Joseph Black or Alessandro Volta. New business forms, pioneered by technologists like James Watt or entrepreneurs like Arkwright, dovetailed with original theories from the likes of Adam Smith. They are all connected – ideas in science and philosophy, politics and technology, directed and undirected, lucky accidents and long programmes, all co-evolving and informing one another.[8] Whether you were a grand *philosophe* at an aristocratic Parisian salon or a sooty-handed Derbyshire engineer, you could be part of a revolution in ideas, built from ideas.

What's more, it's clear that the set of ideas behind the Great Divergence were of unusual consequence. Some were utterly practical, like the flying shuttle or the seed drill, others more abstract, like Newtonian mechanics. Some originated in one realm but led to another – steam engines were the product of tinkerers, built without exact underlying scientific principles, but their invention and refinement later led to an understanding of thermodynamics. Some big ideas related to concepts of political freedom or economic organisation, some to art and entertainment. Sometimes change was gradual, by increments that evolved thanks to countless artisan experimenters. But it could also happen quite suddenly, or seemingly so – the discovery of Uranus or the premiere of Beethoven's Ninth, even if their genesis was longer and richer than that implied. And in this complex, sprawling picture of change and ideation, the accelerating onrush of ideas refashioned the human frontier at a rate never seen before.

Understanding history means understanding both ideas and the conditions that give rise to ideas. They are not the decorative flowers gilding human civilisation, outgrowths of everything else; they are the soil, the atmosphere, the supporting superstructure itself. In the words of the novelist Victor Hugo, 'There is one thing stronger than all the armies in the world, and that is an Idea whose time has come.' Big ideas matter. Having an account of how ideas form, mutate, spread, combine and recombine and grow, and what they are in the first place, matters as well.

Ideas shape and push back the frontier. Some do so more than others: ideas, like Archimedes' or those of the Industrial Enlightenment described above, spark new fields, paradigm shifts, revolutions of all kinds. We need to define those ideas and examine how they work in practice. What is it about big ideas that makes them different?

Let's start by considering how to understand a 'big idea'. Put bluntly, these ideas have the biggest impact. We can venture an initial definition along the lines:

> *A big idea creates a new space of action or understanding for the human species, and those actions or understandings are adopted to the extent that they ultimately affect the life of the average person in a developed context.*[9]

More weakly, but perhaps more workably at most scales of analysis, I argue that:

> *A breakthrough idea creates a new space of action or understanding in a given field or subfields, and those*

actions or understandings are adopted to the extent that they ultimately affect the practice or knowledge of the average practitioner in that field or subfield.

This can be tested relatively easily. Large datasets like the biomedical database PubMed or the US patent record let researchers analyse millions of journal articles and patents, enabling them to see which patents and articles are, for example, most cited, or correlated with the biggest rise in market values. Take any field, then, and if you can assemble a workable dataset, you can use it to see which ideas are most influential, either within a given time period or overall. To fulfil the above definitions, it's likely that less than 5 per cent, potentially much less, of the ideas in any given field would qualify. And while any divide is artificial, given that ideas exist along a continuum of impact, it's obvious that some ideas cluster at the upper end of that continuum. Those are the ideas I focus on here.

It also implies that big ideas are likely to be found 'upstream', in more fundamental technologies or areas of research. The more you burrow down, the less overall impact will be seen. Big ideas work at the foundations and ripple out, moving through a diversity of fields, changing all of them.[10]

But clearly that isn't quite enough. The first part of the definition also needs attention. There are plenty of things which impact everyone, but they don't capture what I mean by a big idea. Changing the packaging of your shampoo or the design of your word processing software may touch a lot of people, but it doesn't qualify them as big ideas. Influence is a necessary, not a sufficient condition. We also need to isolate the sense of originality that should accompany a bona fide big idea.

Science, knowledge, technology, art, business and politics

don't always proceed with linear grace; instead periods of stasis are followed by eruptions of activity, fecund golden ages that glitter in the historical record. They are moments rich in big ideas, and good sources of inspiration in nailing down what those ideas are and how they work.

Speaking of disjunctive leaps inevitably brings us to Thomas Kuhn's *The Structure of Scientific Revolutions*, the most cited work of social science in the twentieth century and populariser of the aforementioned term, the paradigm shift. Kuhn was interested in science's most revolutionary moments, in other words its big ideas: Copernicus and his heliocentric model of the solar system; Lavoisier's oxygen theory of combustion; Einstein wrenching physics from Newton's grip.

Kuhn argued that those moments break what he called 'normal science', the kind which fills in gaps, confirms the world picture, forms a steadily growing stockpile of knowledge. Most researchers spend their lives working on such science: finding data, refining measurements, collecting observations, corroborating and articulating theories in greater detail. This work is underwritten by the paradigm, the 'universally recognized scientific achievements that for a time provide model problems and solutions to a community of practitioners'.[11] But Kuhn also argues that the paradigm is broader, forming 'the entire constellation of beliefs, values, techniques' in a given age.[12]

Inevitably this constellation sets limits. Normal science tends to know what it is looking for, and finds it. The radically unexpected is, by definition, almost never the goal. Paradigms have no interest in destroying themselves. They shape how the practitioners of science see the world and relate to their discipline. But eventually, what Kuhn calls 'anomalies', persistent discoveries or questions falling outside the paradigm, begin to tell.

The Ptolemaic astronomical system, dominant since antiquity, had long been known to include inaccuracies. Eventually these became so blatant that someone, ultimately Copernicus, had to propose a convincing alternative. This is the genesis of a crisis, a period of epistemic and professional uncertainty followed by a new, replacement paradigm. Such moments produce revolutionary history-making science, 'a reconstruction of the field from new fundamentals'.[13]

Despite reservations, Kuhn was conscious of his model's wider potential, writing: 'To the extent that the book portrays scientific development as a succession of tradition-bound periods punctuated by non-cumulative breaks, its theses are undoubtedly of wide applicability.'[14]

Many thinkers and writers find echoes of this structure across other fields of endeavour. Stephen Jay Gould and Niles Eldredge's idea of punctuated equilibrium has a similar structure. A revisionist Darwinism, it rocked paleontology, evolutionary biology and the wider scientific community when first published in the early 1970s – itself a potential paradigm shift. Its authors observed that evolution, unfolding over geological time, bears a structural resemblance to Kuhn's account of scientific progress. Instead of seeing species as evolving regularly, Gould and Eldredge argued that speciation worked in sudden bouts of rapid change. Evolution, they said, works in bursts, 'punctuated equilibria', change occurring in windows of around 40,000 years: a blink in deep geological time.[15]

Like Kuhn, Stephen Jay Gould was alive to the model's potential.[16] He cites evidence from the history of human artefacts and culture, tracing the history of writing technologies as one of punctuated equilibria, beginning with the clay tablet and stylus around 2500 BCE and taking us up to the age of ebooks and electronic text. The history of writing is one of static methods and technology, stable for hundreds of years, spliced with moments – like the invention of paper, the codex or the printing press – of massive change.

Inventions are species of human creation. Joel Mokyr divides them into what he calls macro-inventions and micro-inventions.[17] The former are pathbreaking general purpose technologies whose impact is felt across sectors. They are enablers of further swathes of invention, wholesale productivity boosters. Meanwhile micro-inventions centre on everyday product improvements. They are cumulative, local, small scale and occupy the vast majority of what counts as invention, but nonetheless sometimes concatenations of the latter lead to the former.

Many others have suggested similar structures to illustrate how markets, companies, products and technologies form and flourish. Some innovations have outsized impact – Simon Kuznets called them 'epochal innovations' and the economist Carlota Perez explicitly refers to them as 'techno-economic paradigm shifts'.[18] Epochal innovations are crises that initiate new paradigms. The writer and executive Safi Bahcall talks of 'loonshot' business, say a small drug company or indie film crew, versus 'franchises' like big pharma companies and Hollywood studios.[19] Franchise projects are entrenched blockbusters: the last *Harry Potter*, not the first; the iPhone XII, not the iPhone.

Another analogue is what Peter Thiel calls '0-1' businesses or ideas.[20] Most businesses are '1-n'; they simply extrapolate possibilities from the kernel of an existing idea. In contrast, '0-1' ideas exhibit something completely new, a 'vertical' or 'intensive' progress. It means searing originality, not copying or improving; the creation of new technology versus the globalisation of that technology. This kind of true creation, akin to a loonshot or revolutionary science, is hard and, like bursts of speciation, rare.

These models describe how big ideas do something different. At the heart of each is the sigmoid, logistic or S-curve (see Figure 1). An inflection point sparks a wave of rapid change, which eventually forms a new normal. But it starts with a significant, radical, initiatory moment, a rare pivot point like those fixed by Archimedes, at the bottom of the curve.

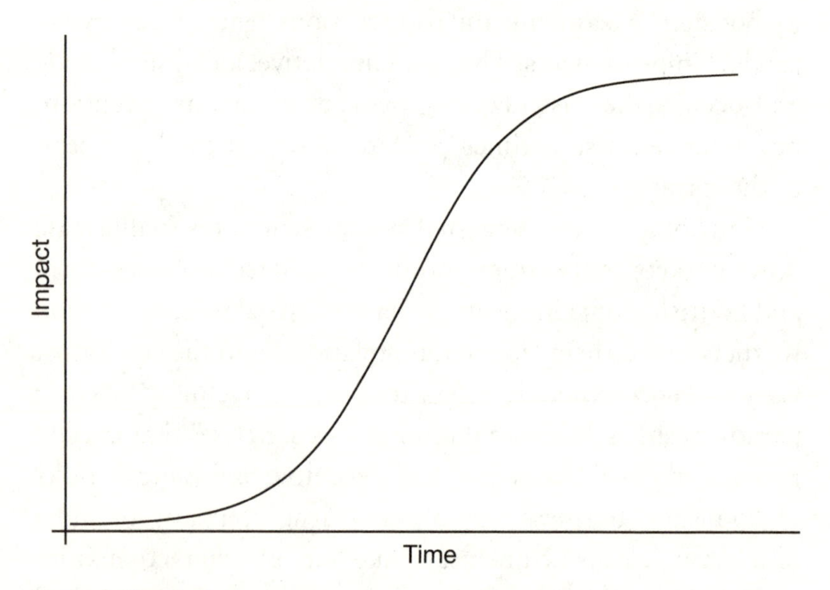

Figure 1: The S-curve

Can these approaches create a concrete working model of what a big idea looks like? A 2019 paper analysed millions of US patents from 1840 to 2010 to find patterns in what the authors call 'breakthrough innovation', 'distinct improvements at the technological frontier'.[21] To do that they needed a way of identifying such innovation. They used two criteria, which follow the approaches taken here: a measure of originality, and a measure of influence. Breakthrough innovations are patents which exhibit the highest scores on both, adjusted for time. Using the text-analysis methods, they compared patents' language, looking for novel uses of language (originality) and echoes of it down the patent record (influence). They then produced indices of patents that isolate major breakthroughs: the telegraph, the telephone, the automobile, the aeroplane, plastics, microprocessors and genetic engineering all show up clearly. The organisations and companies behind the big ideas also look familiar: General Electric and Westinghouse, IBM and RCA, Microsoft and Apple.

These researchers weren't plucking breakthrough innovation out of thin air – they'd found robust statistical means of identifying it. Moreover, similar approaches are widely deployed, suggesting that in almost every field there is scope to apply, *mutatis mutandis*, such models.[22] In science, papers can be analysed in similar fashion.[23] In literary studies or musicology, for example, measures of stylistic innovation and their influence do for authorial experimentation what patent analysis does for technological change.[24] In essence, statistical techniques in almost every field can build a snapshot of which ideas push back the frontier farthest.

But big ideas are also shocks. The frontier has a psychological aspect; what counts as a 'big' idea, what a paradigm shift looks like, exists partly in the eye of the beholder. Many of the most important patents, for example, relate to iterations of the sewing machine; some of the most significant stylistic innovators of the nineteenth- or twentieth-century novel are now forgotten, their language effectively dead. Quantitatively they might even fit the S-curve model and yet something, that Copernican X-factor, is still missing. In this sense the boundaries of big ideas remain unfixed. Rather than deploy an over-rigid scheme, I am interested in identifying an overtly flexible framework; a wide-angle lens, rather than a tight taxonomic grid.

For the most part, we recognise big breakthrough ideas. They cause crises in normal science; they punctuate disciplinary equilibria; they are disruptive innovations, new genres of cultural production, moments of outstanding genius. Not only do they allow us to calculate volume; as we will see, they confirm the germ theory of disease, enable heavier-than-air flight and redefine the nature of music. They are landmarks of enquiry: Newton's *Principia*, Harvey's *Motion*, Lavoisier's *Chemistry*, Lyell's *Geology*, Smith's *Wealth of Nations*, Darwin's *Origin of Species*, Freud's *Interpretation of Dreams*. But they are also the businesses that altered history with a new practice: think of the scale of the Dutch and English East India Companies, industrial techniques in Arkwright's cotton mills or Wedgwood's potteries, major conglomerates like AEG, process and product innovators like Ford, new service models like McKinsey, the mathematical finance of a hedge fund like D. E. Shaw, the digital sprawl of companies like Google and Facebook. Each book or business is not only sharply distinct from what went before, but opens a new space of possibility that defines the subsequent landscape. There are limits to incrementalism: you can improve the horse or cart or embellish Newtonian mechanics all you like, but doing so doesn't deliver the automobile or special relativity.

Big ideas can be found in all areas of human endeavour. You find the above pattern with linguistic philosophy; the Internet; human rights; the concept of zero; the steam engine; the iPhone; utilitarianism; calculus; the periodic table; helicopters; entropy; double-entry bookkeeping; written constitutions; writing itself; deep machine learning techniques; Jacobean tragedy; *Spacewar!* and *Grand Theft Auto*; information theory, quantum theory and game theory; Cartesian grids, rationality and ego. This is an ecumenical approach to ideas, but only by taking such an approach can we see the overarching picture of change, or its absence.

*

There is something undeniably romantic about the notion of ideas as heroically catalytic moments. It's also misleading. It suggests that breakthroughs conform to the Eureka myth; that our most significant instances of thought or endeavour are not rooted in material conditions, or constrained by their histories. Like the original Eureka moment, breakthroughs aren't *ex nihilo* miracles. Like all ideas they are composed of pre-existing ideas recombined. It is not then that breakthroughs are radically new ideas born whole; rather they are especially significant *combinations*, formed by slow processes of accretion at the bottom of an S-curve as much as bolts from the blue.

Although we might associate big ideas with outsized impact, that doesn't imply a speedy or completionist account of their production. When we think of natural selection and evolution, we think of 1859, *On the Origin of Species* and Charles Darwin. In some ways it is the archetypal big idea – a huge, world-changing scientific notion associated with a single author, year and book. But of course Darwin didn't 'invent' evolution. He built on theories from the likes of Anaximander and Lucretius, Erasmus Darwin (his grandfather) and Jean-Baptiste Lamarck. Darwin read Adam Smith, and was thus familiar with the idea that an undirected process with numberless small instances of local competition could have extraordinary results: in Smith's case this was economic growth. Darwin was familiar with Thomas Malthus and his studies of population. Charles Lyell's key work on geology had radically shifted the perception of time. Darwin specifically acknowledged that 'descent with modification' had been recognised by thirty-four predecessors.[25] Indeed, he was a generalist relying on an extensive communication with specialised experts – he maintained a continuous discussion with hundreds of correspondents (at least 231).[26] His research was the work of a lifetime – a gradual realisation over years on the *Beagle* and decades of patient study.

Natural selection is a classic big idea. But its long, complex gestation, its prolific mixture of existing theories and ideas – these are typical as well. Big ideas don't spring into being fully formed, even though it can look like that. Rather, ideas themselves form and grow in an evolutionary manner. Discrete leaps are, under the hood, often the results of gradual cumulative processes and unlikely admixtures. Big ideas 'broker' other ideas in interesting ways – whether that's Elvis Presley brokering gospel and the blues or Gutenberg's printing press coupling the wine press with the idea of casting a seal.[27] Johannes Kepler united the previously disparate fields of physics and astronomy, using new data uncovered by Tycho Brahe to prove the elliptical orbits of the planets. Marx's theory of capitalism combined Hegel's philosophy with classical political economy and an emergent socialist tradition. Likewise Freud didn't 'invent' the unconscious, an idea that had pedigree everywhere from the Upanishads to Thomas Aquinas, Montaigne and Romantic artists. Picasso exploded Western traditions of art by bringing in new, supposedly 'primitive' forms from Africa and elsewhere. The Wright brothers combined the aerodynamics of bird flight with bicycle technology. And on and on.

Zoom into any given area and there is what the writer Arthur Koestler called 'bisociation'; collisions between previously unconnected ideas.[28] Ideas are fusions, productive confrontations of older ideas. Every 'new' is a new synthesis.[29] This means that the prior nature, structure, communicative architecture and social context of ideas powerfully conditions what comes next. To understand the future of ideas means to understand clearly their past and present.

This aggregative sense takes us away from Eureka, but we

need to go further. In that story the critical moment happens in the bath as a sudden, singular moment of realisation. It forgets that Archimedes would still have needed to conduct experiments, double check results. It omits that he would have to think carefully about presenting his findings. After all, he dealt with an all-too-human court riddled with power politics and petty jealousy. And it doesn't say how Archimedes' insight became canon, part of general human knowledge, a staple of the global curriculum. Someone codified it, someone taught it. Ideas don't spread or enact themselves.

Hence breakthrough ideas actually involve several discrete stages. Every idea must go through:

Conception: Archimedes in the bath. Newton watching the apple tree. The spark, the initial cross-fertilisation. Conceptual ignition. Often this is surprisingly slow; it took Darwin decades to formulate his theory of natural selection after an initial suspicion. Epiphanies can come in dribs and drabs as well as in sudden spurts.

Execution: Darwin not only had to think through his ideas, eventually he had to publish them. The execution is the initial paper or book, the proof of concept, the prototype, the unveiling. It is how an idea is enacted, demonstrated, shown to the world. If a breakthrough stays in your head, it's not a breakthrough.

Purchase: *On the Origin of Species* didn't win everyone over immediately. (Indeed, it still hasn't.) It needed debate and scrutiny, elaboration, further proof, serious discussion and active persuasion. But eventually everyone working in biology, arguably everyone with an exposure to science, was influenced: it had gained purchase. Purchase describes the point of widespread diffusion, adoption or acceptance wherein a breakthrough can fulfil my conditions of impact.

It may be that in some eras one stage of the process becomes easier, while blockages accumulate in others. Different ideas might struggle at different points. Some papers are written in a blizzard from concept to publication, but then languish for decades or even centuries before being uncovered and finding purchase. Grasping the future of big ideas means finding the contemporary pinch points on this spectrum.

Big ideas are fragile, imbricated with forces far beyond the control of any individual or even any society. Two of those forces are particularly indicative of how ideas function. The first is luck. In the annals of invention, discovery and creation, the role played by serendipity is dizzying. Robert Koch created bacterial cultures after accidentally leaving a potato out to go mouldy, while a few years later Alexander Fleming stumbled on penicillin by accidentally leaving such a culture in his laboratory sink during a spell of freak weather.[30] Radiation and X-rays were both uncovered during the search for other things. Columbus found the 'New World' by mistake. The pacemaker was meant to record the human heartbeat, not control it. Happy accidents are behind inventions from Newcomen's steam engine to the spinning jenny to vulcanised rubber. Just as every idea is formed of other ideas, so it also involves an element of chance – a random meeting of minds, a lucky experiment, a date missed, an accidental find, a serendipitous connection.[31] Misreadings, faulty copies and fluky mistakes are legion, as powerful, if not more so, than directed efforts or 'heroic genius'.[32]

At the same time ideas repeatedly exhibit multiple discovery, wherein many researchers stumble on the same breakthrough at once.[33] Examples include calculus, oxygen, logarithms, evolutionary theory, photography, the conservation of energy, the telephone and the polio vaccine. No fewer than twenty-three people can claim credit for the invention of the lightbulb.[34] In the early 1920s two researchers, William Ogburn and Dorothy Thomas, found 148 examples of multiple discovery in science alone – and this barely scratches the surface.[35] Artistic movements sometimes seem to spring onto the public stage with styles and coteries fully formed. Polygenesis, in seeming contradiction to luck, suggests that individual endeavours and chance events don't matter that much.

Chance plays a huge part in the specifics of any given breakthrough, but they also have their time. Arthur Koestler called this phenomenon 'ripeness', suggesting that a given society must be ready for an idea. He cites the steam engine, invented as a mechanical toy by Hero of Alexandria in Egypt during the first century CE but only fully realised in eighteenth-century Britain.

Look closely at any breakthrough, and you find it rests on a mind-boggling series of contingencies; zoom out and they look something like an historical inevitability. In fact, the latter creates conditions for the former; the wide context creates the inflammable matter for quasi-random sparks. Both tendencies indicate that big ideas are acutely sensitive to contexts and trajectories, both on the micro level of encounters, conversations, the weather, and at the macro level of economies, cultures, intellectual history. This means that we, just as much as Archimedes and Hero of Alexandria, are constrained by the soil in which we work. Some ideas are likely to be made 'ripe' by our society, others blocked or ignited by random chance.

What, then, of the ideas of the twenty-first century and beyond? We know an exciting set of transformational ideas are being conceived, executed, finding purchase. We know they will be composed of prior ideas, but will nonetheless stand out. We know too that they will be grounded in societal context, but that luck will play a role.

The question becomes whether we are creating the right conditions for their continued evolution today. Before turning to that question, some caveats.

Caveat Ideator

Human frontier, big ideas – these admittedly grandiose sounding notions naturally prompt questions. First comes the scope of the task. Looking at the human frontier and big ideas in general is clearly quite an undertaking; one impossible for any individual, department or even institution. Which is why I make not the slightest claim to comprehensiveness. The aim of the book is to go wide, to connect and synthesise disparate insights from usually disjointed fields and perspectives, while eschewing any claim to completeness. It doesn't aim to cover every possible base – a hopeless task – but rather to step back and connect isolated fields. All that does mean sacrifices, omissions and glosses.

The other option, rather than tackle a fundamental question of our times, is to ignore it. That would be a mistake. This is a conversation that needs having. It's worth looking for the overall picture; worth asking challenging and open questions, like what kind of social environments give rise to or block ideas, what incentives and institutions shape them, and why there might be a slowdown when so many feel a sense of bewildering acceleration.

Likewise you might not share my definition of a big idea, which is deliberately loose, or the framing of the frontier. But even if you disagree on the particulars, I hope we can agree that history doesn't make sense without some gradation in the nature and impact of ideas. The closer you zoom in, the more complex the picture always looks. But to deny that some ideas push back our space of operation more than others feels like a basic misrecognition of how history unfolded.

Another objection asks if we really need or want big ideas. In fact, we might wonder if they can often be harmful. Breakthroughs at the frontier are not wholly positive. Wars are responsible for many of our most significant ideas, from metallurgical techniques to strategic understanding, anatomy to the rollout of penicillin, votes for women to the welfare state. Likewise theories often unsettle, from the great decentrings of Copernicus or Darwin to the destabilisations prompted by Heisenberg and Gödel or the assault on established knowledge launched by critical theorists. New technologies can disrupt and destroy jobs, polities and culture. Technology comes with biteback: industrialisation led to climate change; DDT collapsed ecosystems; CFCs punched a hole in the ozone layer; communism was supposed to improve life for the masses, not a tiny cosseted elite; the cinema and television became a tool for totalitarian propaganda; social media did not bring people together so much as firm up their hostile tribes. Arguably the problem with Silicon Valley is its coupling of an infatuation with big ideas with the failure to grapple with their consequences. Moreover, especially today, technologies can introduce systemic vulnerabilities and existential risks; create new ethical alarms or shocks to the world order; exacerbate and produce inequalities.

Perhaps, given the risks and the damage done, we should just let the frontier rest. Maybe the whole notion of big ideas is a bit dated, passé, arrogant. It could be that we need small, context-specific, grassroots approaches instead; after all, there is value and beauty in the modest and small scale.

To that last point, they needn't be in opposition. Things

like hyper-local economic or political forms or concepts like frugal innovation are big ideas with the patina of the small. And of course, throughout history, as we have seen, many big ideas from the steam engine to evolution are as much the concatenation of smaller ideas as fully formed Big Ideas. Furthermore, in many (but certainly not all) areas I think there is an intrinsic reason to chase these ideas; in science, philosophy or aesthetics for example, in fundamental enquiries and universal questions, debating and searching for big ideas is innately worthwhile. From our earliest days we have been captivated by the search for deep explanations, for mastery of our physical environment. It is part of who we are.

Humanity also needs big ideas. In his book *Collapse*, Jared Diamond narrates how a series of island peoples saw their civilisation fall apart, ultimately to be extinguished for ever.[36] Stuck on their own, the people of Easter Island or tiny Pitcairn Island, or the Norse in Greenland, found what they thought was a sustainable lifestyle, often for hundreds of years at a stretch, and managed to thrive at the very margins of human life. But problems accumulated. Weather changed for the worse. Crops failed. Violence spiked. Soil was degraded and eroded; resources were exploited to the point of diminishing returns before falling apart. The islanders became vulnerable to external shocks; meanwhile they were culturally trapped, unable to fully recognise or grapple with a tidal wave of problems. It wasn't that they weren't smart or didn't see the issues that were building; it was that they were bereft of solutions. Alone, they had no answers, no new ideas.

We are different to those peoples in every respect bar one; on the grandest level we too are an island. This is what remains so haunting about those collapsed civilisations: in cosmic terms, planet Earth is just a much larger version of Easter Island. Sure, we have more awareness, resource, technology and knowledge, but as on Easter Island no one is coming to save us. It's not hard to imagine the long-emptied skyscrapers of New York or Hong Kong piercing the lapping ocean like so many Mayan temples over the jungle canopy. Nor are Easter Island style 'ecocides' the only means of civilisational collapse: disease, wars, raids, elite ossification can all do the same. In 146 BCE the Romans utterly obliterated Carthage, even salting the ground to ensure it would never come back. Of course, in time they fell prey to their own contradictions and the 'barbarians' at the gate.

Now think of the panoply of risks facing the planet. Global climate breakdown. Solar flares. Bioweapons, nuclear weapons, cyber weapons. Antibiotic resistance. Rogue AI. Biodiversity collapse. Resource depletion, from fresh water to rare earth metals to fertile topsoil. Dementia and the ageing of the world. Another major, more severe pandemic and the fallout. Stalling economic growth (and some might say, conversely, continuing economic growth). Political and ideological pathologies. Cultural stagnation and xenophobia. Frankly it may be quite rational to be terrified, and while some if not all of these problems are the fruits of previous ideas, the result of moving the frontier, we still need answers.

In her epic study of flawed decision-making, the historian Barbara Tuchman argued that a large part of the reason why, when faced with catastrophic problems, nations, peoples and individuals make bad decisions is 'mental standstill or stagnation'.[37] Even for much larger and better connected civilisations than those of the lost Pacific islanders, problems start to overwhelm when culture and ideas calcify, when governing elites stupefy in their own dated models. To grapple with novel problems, she suggests, societies need novel thinking. We are not the Minoans of Crete or the Khmers of Angkor Wat, but how arrogant, how dangerous, to pretend that we can blithely avoid mental stagnation in the face of our challenges. In the words of Joel Mokyr, 'while technological progress is never riskless, the risks of stasis are far more troubling. Getting off the roller coaster mid-ride is not an option.'[38]

At present levels of social and technological development, the world cannot sustainably support a population of many billions. Expecting either those in rich countries to give up their way of life or those in developing countries to abandon their ambition feels, at best, misguided. Try telling a family in rural India that the car, fridge and microwave they've been saving for are no longer feasible. Only pushing back the frontier, not just in technology, but across all dimensions, can provide genuine answers to this series of unprecedented problems.

Ideas are not straightforwardly 'good'. But experience suggests they enable us to overcome our challenges, from Malthusian collapses to zoonotic pandemics. Ignoring or actively stifling the most impactful ideas is, even more than pursuing them, a recipe for disaster.

The great Polish science fiction writer Stanisław Lem, who foresaw the Internet, virtual reality, genetic engineering and nanotechnology, quipped that nothing ages as fast as predictions about the future. 'Arrow-flight' projections which simplistically extrapolate present trends invariably get mugged by reality. Expert forecasters are routinely and egregiously wrong.[39] 'Futurology' attracts widespread and often justified suspicion. Gazing into the future is fraught with issues. Meanwhile the future of big ideas is unbounded, dynamic and fuzzy. Hence it is not a probabilistic puzzle but a mystery, or rather a set of interlinked mysteries, all subject to 'radical uncertainty'.[40] In Isaac Asimov's Foundation series the discipline of psychohistory can model the future with mathematical precision. We remain some way off, to say the least.

And yet Asimov and Lem did guess at much that was valuable, not least recent debates on AI and genetic experiment. An author like H.G. Wells was often wrong, but still predicted the atom bomb. Speculative writers pose questions that have deep value for understanding the present and how it conditions the future.

This book is written in that spirit – one using the future as a lens to understand the present, and vice versa. It doesn't claim to have a complete picture of the world in 2100. But it does aim to describe the forces that will take us there.

In the rest of Part I therefore I take up the Great Stagnation debate. I ask why so many people believe we are getting stuck at the frontier, and consider what it might mean to say that when so much around us seems to be changing, when ideas appear so plentiful. Let's begin by looking at an area that concerns us all: medicine.

2

The Breakthrough Problem

Matters of Life and Death

Summer, 1879

Louis Pasteur, aged fifty-seven, already the most feted scientist of his age, was on the cusp of a new breakthrough. Pasteur had been studying chicken cholera. While preparing the bacillus, he accidentally left the cultures in his laboratory for the summer. Returning in the autumn, Pasteur stumbled across his old experiment. Picking up the research, he injected a group of chickens with the old bacillus. Unexpectedly they didn't become severely ill, but actually recovered. Pasteur assumed the cultures had somehow gone off and tried again, injecting those same chickens and a new set with a fresh batch of the disease.

Then something interesting happened. The new chickens died; but the chickens previously injected with the old culture once again survived. Why would those chickens – against all reasonable expectations – live on? Upon hearing the news Pasteur fell silent, then 'exclaimed as if he had a vision "Don't you see, these birds have been *vaccinated*!"'[1]

Vaccination had been known of since at least the late eighteenth century, when Edward Jenner realised that cowpox could create immunity to smallpox, a devastating killer.[2] But until Pasteur no one had generalised from there to form a foundational medical principle. He saw the link between his spoilt culture, cowpox and immunity. Despite everyone knowing about vaccination, it was only he, at this moment, who made the decisive breakthrough. 'Fortune favours the prepared mind' is one of Pasteur's most famous quotes. Few if any minds were as prepared as his.

Leaving a culture suddenly held out endless promise; here was, potentially, a powerful mechanism for the treatment of disease. As he said in a speech presenting his results, 'We can diminish the microbe's virulence by changing the mode of culturing.'[3] At a stroke, vaccines might be actively produced in the laboratory. Vaccination had first been noticed by chance. But Pasteur, with mounting excitement that gave him sleepless nights, saw the possibility of directing that process.

Experimenting with anthrax, he and his team realised that weakened versions of the bacteria produced subsequent generations that were also weakened. In February 1881, he announced his results at the Académie des sciences: anthrax, a terrible livestock disease, one of the biblical plagues of Egypt, was controllable. Defeatable. High-profile public trials in the face of fierce scepticism, always a speciality of Pasteur, once again proved him right; as the tests got under way near Melun, France watched with bated breath to see whether or not the vaccinated animals would die.

Having injected vaccinated farm animals with a lethal dose of anthrax he would have to wait two days for the results. His critics gazed on forensically. Jealous of his success and sceptical of his theory, they wanted nothing more than to finish Pasteur's ideas and his glittering career alike. It looked touch and go . . . But two days later the infected sheep were still alive. Pasteur had triumphed, and so had vaccination. The field and practice of immunology was born; hundreds of millions of lives would, in time, be saved.

From here Pasteur went on to develop an anti-rabies vaccine, a project of personal importance: he remembered a rabid wolf ravaging his childhood hometown, causing the deaths of eight people. To do this he worked with microbes that were longer acting and more difficult to find: viruses. Even here Pasteur found a way to build immunity, despite being unable to directly see the rabies virus. In the first trial he vaccinated a young Alsatian boy, Joseph Meister, who had been bitten by a rabid dog. The use of a human test subject was far ahead of schedule, but events overtook Pasteur as the clock ticked down on the boy's life.

It was still an awful risk. One long-time assistant, Emile Roux, walked out of the lab in protest (he later returned, and went on to develop the diphtheria vaccine). But Pasteur knew that doing nothing was a death sentence. He had surmised that the vaccination would work ahead of the month-long incubation period. For it to do so, he would need to inject the virus before any symptoms appeared. With a feeling of dread, Pasteur began administering small doses of rabies in the full knowledge he could be making the situation worse. But, after twelve rounds and weeks of sleepless nights, success.

The boy grew up to be the gatekeeper of the Institut Pasteur in Paris, until he was killed by the Nazis for refusing to open Pasteur's grave. Meister was the first known human to survive rabies. Infectious disease had reigned almost unchallenged over humanity since the dawn of time. No longer.

Had Pasteur developed vaccination alone, we would still remember him as a giant of medical science. But this was just the culmination of a series of breakthroughs without which the modern world would be inconceivable.

Pasteur was born in 1822 in the town of Arbois, in the Jura; his father had fought under Napoleon at Waterloo. A dogged and talented student, Pasteur became interested in the discipline of organic chemistry. By 1854 he was appointed dean of a new, practically focused science academy in Lille. As the youngest person to occupy such a position in France, he threw himself into the project. His was to be applied science solving real-world problems. One of these involved fermentation. A local factory owner, who made alcohol from sugar beet, was having difficulties. The products of his fermentation kept going bad. At the time people didn't fully understand fermentation: was it a purely chemical or a biological process?

Pasteur spent weeks at the factory in careful empirical study, watching and talking to the workers, collecting beet sugar samples. His years using a microscope paid off when he observed something remarkable: yeast reproducing itself. It was clear evidence that yeast was alive; that fermentation was biological. Microbiology came of age.

Further experiments followed. Pasteur began to understand that milk went sour thanks to the operation of certain micro-organisms. A minute and hidden world, fleetingly glimpsed only since the seventeenth century, began to reveal itself. Pasteur started to wonder if these creatures might be responsible for disease.

Following that insight, he turned to a huge scientific controversy of the day: spontaneous generation. His experiments decisively showed that the micro-organisms which became visible under magnification were carried on the air; they didn't, as many believed, spontaneously generate. It was a revolutionary idea, bitterly contested.

Pasteur was always attentive to public communication,

giving high-profile lectures at the Sorbonne to the great and the good. His fame was growing from around 1860 and he even had an audience with the Emperor Napoleon III, who tasked him with finding out why wine spoiled. Back at Arbois, the investigation again involved a long, hands-on process working with farmers and samples; it led to the technique of pasteurisation, a gentle heating that arrested the microbial growth responsible for spoiling, not only in wine but also milk, cider and cheese. Instead of keeping the patents, Pasteur made them public. Thanks to pasteurisation, which is still fundamental to safe food supply, millions of lives would again be saved.

He also investigated a plague that was killing European silkworms. Over years he watched the silkworms, his experiments revealing ways to identify the disease involved. It all provided more evidence of his crucial germ theory, confirming that the microbes were responsible for illnesses and so suggesting an answer to one of the most pressing riddles of human life: the nature of infectious disease.

War with Prussia in 1870 meant Pasteur was required to confront how germs get into open wounds. He urged a cleaner approach to field hospitals. Yet thanks to his lack of a medical background doctors often simply didn't believe germ theory. The food, the weather, bad luck: all were blamed for the atrocious hospital mortality rates. This same process was repeated in maternity units: many women died thanks to doctors' gore-sodden and infected hands.

There were profound barriers to progress, not least poor tools and technology, a lack of wider understanding and the professional hostility of surgeons, clinicians and veterinarians, whose entrenched resistance to, for example, the cleaning of operating theatres, provided bitter opposition to his advances. Doctors with pet theories of disease didn't want to know about germs. Chemists did not want to believe fermentation was biological.

Pasteur worked on his ideas in a world far removed from the well-funded, capable, technologically advanced labs of today. Despite being director of scientific studies, he worked in small, dingy rooms, starting in the attic, often building or buying his own basic laboratory equipment, and assisted at first only by his wife Marie. Later, at laboratories in Arbois and elsewhere, he used similarly self-designed and jerry-built equipment. It was only on the repeated orders of Napoleon III that Pasteur was provided with a serious, up-to-date laboratory. Moreover he became half paralysed after suffering a blood clot in the brain, impeding his own ability to directly conduct experiments.

Since the dawn of humanity, sickness was both a fact of life and mystery. Pasteur changed that. The germ theory of disease, the technique of pasteurisation, an understanding of sepsis and clinical cleanliness, the technique of vaccination, applied to rabies and anthrax, the whole universe of micro-organisms and their myriad interplays: it adds up to a legacy of breakthrough after breakthrough, formed in spartan conditions with rudimentary equipment as their progenitor shuffled back and forth between practical problems and high science. Yet these big ideas transformed the human frontiers of knowledge, medicine, health and even morality.

Pasteur exemplifies a model we have come to take for granted: great progress improves both our knowledge and our technology, initiating a virtuous circle. But knowledge and technology are always in an arms race against the problems they face. Pasteur gave us a decisive lead.

And yet, how many Pasteurs are working today? That is to say, not how many people are working on medical research or microbiology, but what work therein has or could have equivalent impact? One view suggests that Pasteur stands at the beginning of a generalised increase in the production of ideas. But another argues something else: yes, we have an increase, but within it there are fewer ideas with the significance of Pasteur's. Thinking may have become easier, but thinking big is as challenging as ever.

Between the late nineteenth century and the present, human life expectancy underwent a revolution, underpinned by a series of astonishing advances in medical science and public health. The first real pharmaceutical product, the drug Salvarsan, based on a compound synthesised in 1907, offered a cure for that old scourge syphilis. Three and a half decades later came an even bigger breakthrough: the discovery of penicillin and the age of antibiotics and mass medicine.

Although Alexander Fleming made his famous discovery of a bacterium-killing mould at St Mary's Hospital, London, in 1928, it was 1941 before the first person, a policeman called Albert Alexander, was treated with penicillin after a scratch from a rose bush became badly infected.[4] Alexander's experience is a stark reminder of how easily infection used to kill. And, yes, he died, but only because the supply of penicillin ran out. As researchers in Europe began to understand the potential of antibiotics, the action had moved to Oxford where two scientists, the Australian Howard Florey and the German émigré Ernst Chain, began following up Fleming's findings. By June of 1941 Florey was on his way to America where, now backed by four major American drug companies, the rollout of penicillin could begin. In 1945, as the war came to an end, Fleming, Florey and Chain shared the Nobel Prize in Medicine.

Pasteur had started a cycle. For the first time, step changes in medical capability were regular, even expected. We entered a 'golden age' of medicine. In the words of one commentator,

'Medicine, which for most of its history had very limited powers, was quite suddenly marvellously, miraculously effective. There was a golden age of about fifty years, from the mid-1930s to the mid-1980s, when almost anything seemed possible.'[5] Thanks to the discoveries of this period we can kill bacteria and conduct open-heart surgery, transplant organs and produce babies in vitro, regulate pregnancy with a pill and keep people alive on the brink of death in intensive care. And we can eliminate diseases from polio to smallpox.[6]

At the same time, life expectancy, which had remained roughly stagnant for most of human history, improved. Medical progress played a role, but another big idea – public health policy – also came to the fore. Epidemiologist Thomas McKeown gave a classic demonstration of this in the late 1970s. He noted that in the UK, the number of cases of tuberculosis of the lung fell from 4000 per million in 1838 to essentially none per million in 1960.[7] The drugs used for fighting tuberculosis were among the earliest antibiotics, introduced by 1945. But incidence of the disease had already fallen to 350 cases per million by the time they were made available. Ninety-two per cent of the drop in tuberculosis cases occurred before the direct application of medical science.

Mass public health improvements were key, particularly the establishment of an urban sanitation infrastructure. Private indoor toilets were available to only 10–20 per cent of the population in the 1910s; by 1940 over 60 per cent of Americans had them.[8] Running water was adopted even faster, and a majority of the population had it by the end of the 1920s.[9] Europe was even further ahead, completing its journey to sanitation by the late nineteenth century.

Pasteur's germ theory not only combated disease, but meant considerations like food supply became public health issues. The transition to cars took horses, and their dung, off the streets. Hospitals multiplied in number (from around 120 modern-style hospitals in the US in 1870 to 6000 by 1920) and became clean.[10] Doctors grew more knowledgeable, drugs entered the market, and regulated, longer-life items like canned food altered patterns of consumption. Sanitation, better housing, nutrition, cleaner cities and hospitals, better healthcare, safer streets: it was an extraordinary change.

In 1880 infant deaths in the leading economies matched those of Tudor England at 215 deaths per thousand births, well within the 200–250 deaths per thousand births estimated for early modern society. Seventy years later, in 1950, the number had shrunk to just 27 deaths per thousand births. Life expectancy for white American men increased from 48 to 63 in the forty years 1900–1940, a rate of improvement 'never matched before or since'.[11] According to the UK Office of National Statistics, life expectancy was 44.1 years for men and 47.8 for women in 1890. By 1950 it was 66.4 and 71.5 respectively.[12]

Improvements in life expectancy continued throughout the latter half of the century, if at a markedly slower rate. Whereas previous gains had been concentrated in saving the very young, profoundly and happily changing family life, as the century wore on improvements shifted to the elderly. By 2000 British life expectancy for men was 76 years for men and 80.6 years for women.[13] The rate of progress had roughly halved, but progress there was. Until now.

In the UK, the US, France, Germany and elsewhere, we are seeing the first signs that, for complex reasons, life expectancy is no longer improving. Indeed, the US saw consistent falls between 2015 and 2020, the biggest since 1915–1918, the years of the First World War and the Spanish flu pandemic.[14] In Britain a marked slowdown started in 2011, with no progress being made from 2015.[15] At best, Britons are seeing the slowest improvements since the Second World War. The impact of coronavirus is certain to further revise down these numbers. At the frontier, something is going wrong with Pasteur-style breakthroughs. The drugs don't work; at least, not like they used to.

The discovery of drugs appears to obey a rule christened Eroom's Law. In a nutshell, the number of drugs approved for every billion dollars' worth of research and development (R&D) halves every nine years. This pattern has remained largely consistent for over seventy years.[16] Since 1950, the cost of developing a new drug has risen at least eighty-fold.[17] A Tufts University study suggests that the cost of developing a drug approved by the US Food and Drug Administration (FDA) rose at least thirteen times between 1975 and 2009. By the mid-2000s it was $1.3 billion. Today it stands at above $2.6 billion, although science writer Matthew Herper estimates it as $4 billion.[18] In the 1960s, by contrast, costs per drug developed were around $5 million.[19] Timelines, at least pre-Covid, are likewise extended. Eroom's Law shows that it takes more and more effort and money to develop new drugs. Achieving a pharmaceutical breakthrough is on a trend of increasing difficulty.

Eroom is not a person. Eroom's Law simply reverses the name Moore, as in Moore's Law (the idea that the number of transistors on a chip will double every two years, driving an exponential increase in computational power). If anything epitomises technological optimism it is Moore's Law. Eroom, meanwhile, the deep pattern of pharma, works the other way round. Advances don't compound and get easier: the challenges do.

Even in the 1980s a 'dearth of new drugs' was evident.[20] There was a sense, which has only intensified since, that the golden years had ended, that we were in 'the age of unmet and unrealistic expectations, the age of disappointment'.[21] Drug discovery is concentrated in two areas: rare diseases and chronic conditions like blood pressure. Both offer steady, predictable returns. Serious but common diseases have languished, while the challenge presented by something like the common cold still remains. At the same time, pharmaceutical research has a consistent trend for making losses – which doesn't bode well for the future.[22]

This is all deeply strange. It bucks the basic wisdom that a massive escalation in R&D should see massively escalating returns. In 2014, for the first time, global pharma revenues topped $1 trillion. In 1988 the budget of the UK National Health Service was £23.5 billion;[23] thirty years on it was at least £100 billion higher.[24] Population had increased by 10 million or 20 per cent, yet the health budget had risen 500 per cent. In real terms, funding more than doubled. Spending on medicine increases at 5 per cent a year. And in this general respect (at least) the UK performs better than comparative countries which have seen even steeper healthcare inflation. More prescriptions are also being issued: an increase of 50 per cent in the decade 2006–2016 alone. As with research spending, all of this should show up more clearly in the outcomes.

The trend is despite numerous advances in the underlying scientific and technological toolbox. Over the 1980s and 1990s, combinatorial chemistry saw an 800-fold boost in the number of drug-like molecules that could be synthesised per chemist. Molecule libraries, the basic building blocks of pharmaceutical research, grew vastly. DNA sequencing improved more than a billion-fold from its beginnings in the 1970s. Such advances are bolstered by powerful new fields like computational drug design.[25] Health-related research now consumes 25 per cent of all R&D spending, up from 7 per cent in the 1960s.[26] Science, technology and economics all on the face of it imply that drug discovery should be speeding up and getting cheaper.

Eroom's Law bucks the pattern that began with Pasteur. It suggests a steepening challenge that connects to the slowdown in life expectancy improvements. Every year it takes more money, researchers, time and effort to achieve breakthroughs. Each and every one of us is affected – our families, our friends, our basic quality of life. When it's our turn, or the turn of our loved ones, to lie on the hospital bed, these questions feel all too real. Understanding why progress is so uneven has never been more important.

Nowhere is that truer than in the struggle to defeat cancer. In developed countries, 50 per cent of people will be diagnosed with cancer in their lifetimes; worldwide, over 17 million patients are diagnosed each year, and this figure is expected to rise to 27.5 million by 2040.[27] Nonetheless, until recently oncology had only three main treatments – surgery, radiation therapy and chemotherapy: 'cut, burn and poison'. Many expensive drugs have a bad track record. A study published in the *Annals of Oncology* concluded that of forty-seven drugs funded out of a special NHS funding pool, only eighteen increased survival rates and even then by just three months; the rest basically did nothing, but came with a host of side effects.[28]

But the news here is hopeful. We have, perhaps, a textbook big idea in the form of immunotherapy: treatments promising to revolutionise the attritional 'war on cancer'. Some researchers even compare it to the discovery of penicillin: a turning point that will forever transform the field and change countless lives.[29]

Immunotherapy is based on a sophisticated understanding

of the immune system's molecular biology, homing in on T-cells, a kind of white blood cell. Over the last thirty years researchers have realised that cancer plays tricks with the T-cells, using the immune system's own safety checks against it. Cancer essentially fools the body into not attacking it. If scientists could negate cancer's deceptions, the T-cells (and others) could march into battle unimpeded. Another technique samples someone's T-cells, re-engineers them to attack their specific, personal cancer and then introduces them back into the patient – these cells are called CAR-Ts (chimeric antigen receptor T-cells). They too hold great promise.

When the 2018 Nobel Prize in Physiology or Medicine went to Jim Allison and Tasuku Honjo, two pioneers of immunotherapy, it matched the announcement in 2015 that former US President Jimmy Carter had been subject to experimental immunotherapy for cancer and had beaten the disease. The arrival of immunotherapies suggests that we are moving up the problem ladder, finally addressing more causally complex, biologically protean conditions having already 'solved' simpler conditions.

There is a 'but'. To the outsider it seems like a wonderful breakthrough. In fact, the story is much longer and more difficult than that. Immunotherapy's long and troubled gestation and continuing struggles indicate the challenge of big ideas today. We may be getting there; but the road has been longer and rougher than anyone hoped.

For decades cancer immunotherapy was considered a dead end. Although first mooted in late nineteenth-century New York, the story of immunotherapy is one of missed opportunities and leads not taken. Most scientists considered it absurd that the immune system could fight cancer; they didn't believe cancerous cells would ever be recognised as foreign invaders.

Nonetheless, work went forward. False starts were common.

One resulted in a 1980 *Time* headline that heralded a still unproven immunotherapy as 'penicillin for cancer'. Failing to live up to the hype, it rocked faith in the underlying principles. Despite some stunning data, trial outcomes were uneven. Funders wanted unambiguous results. Even true believers began to wonder.

Meanwhile cancer research ballooned, consuming eye-watering amounts of money. Over the last fifty years, probably no single research endeavour can match it for funds spent. In 1971 Richard Nixon started a 'war on cancer' with the National Cancer Act. When his 'crusade' was launched, a cure was thought to be easily achievable; another cycle in a deep pattern of progress that would naturally follow successful treatments for childhood leukaemia. The researchers even believed it might be accomplished by 1976, just in time for America's 200th anniversary.

Cancer still receives billions upon billions of dollars of research funding every year. The US National Cancer Institute has an annual budget of around $5 billion; Cancer Research UK spends nearly £500 million per year. Some estimates argue that around $500 billion has been spent since Nixon's declaration of war; that is, something like $20,000 for every American who has died of cancer in the last forty years. In 2016 – forty-five years after Nixon – Barack Obama and Joe Biden launched the Cancer Moonshot to finally defeat the disease. Yet although there have been improvements in care, the kind of wholesale leaps in progress found in the medical golden age have not occurred. This is not in any way to diminish the extraordinary work of researchers and their institutions; on the contrary, it highlights the colossal challenge they face.

Getting to the point of a breakthrough required major advances in the understanding of cancer and immunity, and billions of dollars of National Institute of Health funding. The first immunotherapy was approved by the FDA in 1992, but even then it remained a fringe treatment. Until the fundamental mechanisms were understood, no pharmaceutical company would take a meaningful risk. Immunotherapy's poor record and the risk aversion of big pharma meant that getting trials approved was an immense challenge.

Meanwhile science had become such a large and disparate field that simply keeping up was difficult; an advance in one area no longer automatically translated into others. Prestigious journals wouldn't publish on immunotherapy for the same reasons pharma companies avoided it: too many failed attempts, the patina of quack science. While the NIH and others continued funding immunotherapy at the margins, other avenues were prioritised.

The point is that immunotherapy is no sudden breakthrough. Like other success stories such as mRNA vaccines, it has taken decades upon decades of blind alleys, missed opportunities, failed careers and cranks grinding away at the margins of science, not to mention, in total, truly monumental amounts of research funding and effort. Compare this idea to those of Pasteur, who worked in a basic lab with a couple of assistants. Fleming, Florey and Chain needed a university department and a research hospital; cancer has required tens of thousands of researchers spread across the world's cutting-edge biomedical research centres.

And we're still not there. Talk to those close to the research and they mention that the results of clinical trials are patchy: immunotherapy seems to work for some cancers and patients but not others. Doctors on the front line are often less excited than the companies developing drugs. And although over two thousand immunotherapies were in trials or the preclinical phase as of 2019, this proliferation creates a new problem:

there won't be room for all those therapies on the market, and the investment boom could once again turn to bust. Immunotherapy prices, moreover, are astronomical: the best-known examples usually cost hundreds of thousands of dollars. Novartis' CAR-T therapy costs $475,000 per patient. In the short to medium term, it is debatable how widespread a cure it can become. Yes, immunotherapy is hugely significant, an attack on cancer and the medical frontier. But to pretend there aren't problems, to ignore its attritional gestation, fails to recognise how medical breakthroughs happen today.

The advent of cancer immunotherapy is truly welcome and inspiring. But it doesn't buck the pattern. It describes the pattern. It isn't an exception to the breakthrough problem; it is part of it.

From Pasteur to Pfizer, the unparalleled arc of medical and public health advance in the last two centuries has taught us to expect miracles. Quietly, however, these have been getting more challenging. This is not to denigrate figures like Pasteur or the extreme difficulties they faced. After all, in the face of ignorance, scarce resource, poor tools and little theory he arguably went further and faster than anyone before or since. That's the point. Somewhere out there is another Pasteur; probably many, many Pasteurs. But it's inconceivable that they alone could have the equivalent impact, despite having much better conditions, bigger teams, more knowledge and insanely improved tools.

Eroom's Law is far from the only example. We face a world where the remaining problems – and the new ones – are of a higher order. At a certain point, endeavours hit a breakthrough problem, where despite the improved capacity for making big new ideas happen, they don't.

There is nothing inevitable about a future rich with big ideas.

From A to B: Our Surprisingly Stationary Transport

17 December 1903; Kitty Hawk, North Carolina

There was too much wind. Gusts of just 30 mph were potentially fatal. After all, only a few years earlier the great aviation pioneer Otto Lilienthal had died after crashing in similar circumstances.

But no matter! After months of delays, technical problems, journeys back and forth with spare equipment; after years of tests, prototypes, research, days and nights spent hunched over workbenches; after decades of dreaming – this was the day, on the Outer Banks, spits of storm-blasted sand sticking out into the hostile Atlantic.

Humanity had imagined flight for millennia. Now the brothers Wilbur and Orville Wright believed it would finally happen. Against the odds they had built a flying machine.

The brothers cut distinctive figures in their uniform of dark suits and starched collars.[30] It was the turn of the younger brother, Orville, to fly. The two men shook hands before Orville clambered into what they called the Flyer. A camera was positioned towards the starting run. One of the onlookers, John T. Daniels, was asked to trigger the camera at takeoff. Orville lay on his stomach, motionless in the centre of the Flyer while the engines warmed up, a process that took minutes. Wilbur held the aircraft's starboard wing, steadying the plane. Despite the dangers the brothers were, as usual, diligent and calm.

At 10.35 Orville loosed the rope holding the Flyer. Slowly it started down the launching ramp, a trolley along a wooden path. Wilbur moved with the plane, keeping it level as it lurched forward into the headwind. Time stood still as the plane reached the end of the launch track. With the observers' hearts in their mouths, the Flyer fitfully took off. It was, as Orville said, 'extremely erratic'. As it launched, Daniels squeezed the rubber trigger of the camera.

The Flyer was airborne!

It bounced wildly, bucking and writhing. It felt dangerously out of control. Orville hung on for dear life. Every time he twitched the rudder it 'overreacted'. As he put it, 'the machine would rise suddenly to about 10 ft, and then as suddenly, on turning the rudder, dart for the ground.' The small group of onlookers watched, amazed and on tenterhooks. Just 120 feet from the launch track, after only twelve seconds, there was an even sharper dart towards the ground, a wing clipped the sand and the Flyer came down.

History had been made. As Orville later said, 'It was only a flight of twelve seconds, and it was an uncertain, wavy, creeping sort of flight at best, but it was a real flight at last.'

At 11 a.m. it was Wilbur's turn. He went further, around 175 feet, with a solid takeoff and clean flying. By the end of the day Wilbur was flying half a mile. They began to understand their control mechanisms despite the frequent and unpredictable gusts of wind. There was almost a last twist. Towards the end of the day a sudden squall sent the Flyer tumbling towards Daniels; 600 lb of machine and canvas caught him. He only managed to escape, miraculously, when the Flyer paused before smashing to smithereens against the sand. All that was left was for Orville to telegraph the family back in Dayton, Ohio. 'Success', his message began.

Between the dunes and wind, these two brothers, amateurs working out of a modest provincial workshop, had executed something extraordinary: heavier-than-air flight.

Afterwards, the idea quickly gained purchase. In 1904 the Wrights embarked on tests closer to home, on Huffman Prairie. The public started seeing the Wright Brothers' wondrous flying machine for the first time. This was now the

Flyer II. By 1905 the brothers were onto another new and improved model: the Flyer III, complete with a 25 horsepower engine, a refined design, a bigger rudder and improved wings.

Over the course of 1905 the Wrights perfected the art of flight, a sensation described by Wilbur as 'a realization of a dream so many persons have had of floating in the air. More than anything else the sensation is one of perfect peace, mingled with the excitement that strains every nerve to the utmost, if you can conceive of such a combination.' That year Orville flew fifteen miles in the Flyer III.

Having applied for the patent in 1903, they were informed that it was granted in May 1906. Even then there was widespread scepticism. But from 1907 the money got serious: a munitions firm called Flint & Company offered half a million dollars for the European sales rights. Before long the German government made a similar offer. In early 1908, after long negotiations, the Wrights had a new deal: $25,000 for a Flyer from the previously uninterested US War Department.

Legendary tests were held at Le Mans and Pau in France, in Rome and Berlin, and in Washington DC, where Orville had a crash that killed his passenger. In New York, the demonstration was capped by a flypast of the Statue of Liberty. The brothers became wealthy and famous. *Le Figaro* was ecstatic about the Le Mans demonstration: 'It was not merely a success, but a triumph ... a decisive victory for aviation, the news of which will revolutionise scientific circles throughout the world.'

Aeroplanes are a big idea par excellence. People have always looked up at birds and wanted that freedom – from the myth of Daedalus and Icarus to the brave experimenter who, in twelfth-century Constantinople, strapped wing-like contraptions to his arms and jumped off the Galata Tower (with unfortunate consequences). Lighter-than-air flight had been cracked in the eighteenth century by the French paper manufacturers, the Montgolfier brothers, who put on the first public display of a hot-air balloon in June 1783. But it wasn't heavier-than-air travel, it wasn't controlled or powered.

The challenge for the Wrights was immense. They had no formal engineering training, few connections, limited resources – the profits from their modest bicycle business – and no support from the government or military. Their first breakthrough came from an imaginative leap: humans needed to copy birds, but not in a literal way. The arched aerofoil shape of a bird's wing was key; its principles could be recreated using a wooden frame strung with muslin. This created lift by having air travel faster over the top, creating pockets of high pressure beneath the wing.

Lift, the brothers soon realised, wasn't the only critical factor: maintaining balance was also essential, keeping the pitch and roll of the aircraft in equilibrium. Close study of birds made Wilbur realise that active balance management was critical in flight. But copying birds wasn't good enough. You had to reimagine them. Wilbur noticed that birds raised or lowered their wing tips in flight, changing the angles at which the wing approached oncoming air. He realised that you could, through such 'wing warping', make a plane turn. Other pioneers wanted power, and saw flight as being like a flying car. This new emphasis on balance was more like imagining a plane as a flying bicycle.

Another problem was trying to measure the lift and drag created by a wing. They responded with characteristic ingenuity, building one of the world's first wind tunnels in their bicycle shop. This led them to develop longer, thinner, flatter wings. Orville had a further flash of imagination while working on aerodynamics. Lying awake one night, he had an idea: why not make the rudder movable? Wilbur liked the idea at once and suggested a further improvement: the pilot should be able to control the rudder with the same mechanism that controlled the wing warping.

In total the Wrights spent about $1000 of their own money on building the Flyer. In contrast, Samuel Pierpont Langley, Director of the Smithsonian Institution, had spent over $70,000 of public money on his 'aerodrome', a lavish flying machine that spectacularly failed to fly anywhere.

In creating the first functional aeroplanes we have the story of a major breakthrough in miniature: the original ideas of human flight, coupled with the deep insights into how it could be done (conception); the building and testing of the Flyer at Kitty Hawk (execution); and then the rapid rollout of trials and production around the world (purchase). At a stroke the potential for human movement was transformed. It was surely almost beyond belief to be one of the first people to witness the Flyer, to feel the impossible made manifest.

Two unknown brothers working with rudimentary tools, and with virtually no capital or assistance, succeeded in changing the course of transport, technology, space and geography; as much as any two individuals, they advanced the frontier of human motion. At the time they were working, transport generally was in the midst of a revolution that saw speeds accelerating and radical new forms of movement rolled out across the world stage. For about 150 years everything continued to accelerate as distances seemed to shrink. But then the process began to stutter. We hit a glass ceiling of transportation.

*

For most of human history, going anywhere had been tiring, uncomfortable, dangerous and slow. Over the centuries roads were not noticeably better than in Roman times. Ships became quicker and cheaper, but developments in maritime technology were incremental.

Then came change. Although better roads, ships and canals all increased speeds, the railway was a truly unprecedented shift: like flight, a textbook breakthrough. The world's first modern steam-driven railway, the Stockton and Darlington, opened in 1825. America's first, the Baltimore and Ohio railroad, started in 1830. Canals were doomed: railways could carry fifty times the freight for the same cost, and an incomparable amount compared to horses.[31] One example gives a sense of how fast and full this transformation was: 'When [President] Andrew Jackson arrived in Washington in 1829, he traveled by horse-drawn carriage, moving at the same speed as the Roman emperors. When he left eight years later he traveled in a train, moving at roughly the same speed as today's presidents when they deign to travel by train.'[32]

In 1869, when Leland Stanford hammered in a final golden spike to unite the Union Pacific and Central Pacific railways at Promontory Summit in the Utah Territory, the US was for the first time linked by rail. At a stroke the transcontinental railroad seemed to change geography, time and space itself: coast-to-coast journeys were reduced in duration from six months to just six days. By the second half of the nineteenth century, thirteen miles of railway track were being laid every single day in the US alone.[33] Smaller countries like the UK or France were, by this time, already stitched together into dense networks.

Everything began moving. There was, for example, a decisive change in ship building. Before steam it used to take around six weeks to make the westbound transatlantic crossing from the English coast to New York. The first steam ships made the trip in 1838, cutting the time to just fifteen days in the case of Brunel's SS *Great Western*. Ten years later Cunard ships could do it in eight days. Seventy years after the first steam crossing the fastest ships could make it in just under five.[34] Steel construction enabled the construction of vastly bigger ships, and powerful and efficient turbine engines could propel them at speed. In the words of Joel Mokyr: 'While the typical ship of 1815 was not much different from the typical ship of 1650, by 1910 both merchant ships and men-of-war had little in common with their steam-operated predecessors half a century earlier.'[35]

Then there was the bicycle, destined to become, at one point, the world's most popular form of transportation. And looming over them all an innovation and idea at least as significant as the Wrights':[36] Karl Benz's Motorwagen, dating to 1885, and based on an engine first built in 1879. The internal combustion engine was integral to revolutionary forms of transport, touching every corner of life from tanks and armoured cars to lorries and tractors. The world of out-of-town shopping centres, abundant food and dispersed families exists thanks to this engine.

From an extortionately expensive curiosity, cars quickly became a mass-produced necessity, manufactured (thanks to Henry Ford's 1907 Model T) by the million. Even as it became normal for American and then European families to own a car, so plane travel went mainstream. By the mid-1930s flying was a viable if expensive mode of public transportation. Aircraft like the Douglas DC-3 could carry twenty-one passengers and cover a thousand miles on a single fuel load, crossing the continental US in three stops and fifteen hours.[37] Improvements in existing systems like roads and shipping, coupled with the dramatic effect of new technologies like railways, motor vehicles and airplanes, gave us a new sensation: acceleration!

Imagine a GI coming home at the end of the Second World War. His journey used transport that seventy-five years earlier did not exist. An American returning home from Europe today will, on the face of it, have a very different experience to that GI: more sophisticated, comfortable, clean and reliable. It would all look bracingly futuristic to a soldier whisked from 1945. But it's still a car or a plane. Seventy-five years on, nothing is completely alien; the journey still consists of perhaps a bus, train or metro journey of some kind, maybe a taxi, a transatlantic flight and then a drive home at the other end. The experience of transportation progress since the war is one of smooth improvement; but before it came a juddering rollercoaster of new transport modes. The transport paradigm did not shift, even if accoutrements and design marched forward.

People often say to me that the world is surely getting faster, that everything is speeding up. But at a very basic level this is not true. Seventy-five years have not enabled us to go much faster.[38] Cars and trucks carry no more people or goods than they did sixty years ago. Vehicle speeds in big cities haven't changed since Victorian times – traffic in London moves at a paltry 7.4 mph, no better than a horse and carriage.[39] Roads are congested, speed limits consistent over time or more stringent, the dimensions of vehicles constrained by the existing infrastructure. In the 1940s it was already possible to buy a car that travelled at 100 mph. You still can today, although as in the 1940s the number of people actually driving at 100 mph is (probably for the best) low.

Jet engines are constantly being improved, but they date back to the 1920s and 30s and the first jet airliner, the de Havilland Comet, began operations in 1951. The first passenger jets crossed the Atlantic in eight hours; it now takes seven, not including the faff at the airport.[40] Supersonic commercial air travel came and went with Concorde. The Boeing 747, the workhorse of the sky's great trunk routes, first flew commercially in 1970 and is still going strong. The fastest planes in history – the legendary Lockheed SR-71 Blackbird and the X-15 rocket jet – date from the 1960s. The Airbus A380 was heralded as a revolution; it's a great plane with countless small improvements. But really it's just a big (and now discontinued) passenger aircraft.

The last time anyone walked on the moon was 1972, after just six lunar missions. The fastest anyone has ever travelled was on the Apollo 10 mission – back in 1969. Bases on the moon and Mars remain the province of science fiction. Humanity goes no further than Yuri Gagarin did. If you showed today's satellites, the International Space Station and atmosphere-grazing space-tourism pods to a child in the 1960s, they would be gravely disappointed.

Our cars, trains, ships and planes are cleaner, sleeker, safer, more reliable.[41] But they do not change the fundamental mode of everyday movement, as did the car, or break the limits of speed, as did flight. Quality goes up, price comes down, as it should in a well-functioning and competitive market. But the progress is incremental, a legacy of plentiful small and welcome improvements that don't add up to big new ideas. And it all comes in the context of industries and R&D centres that are among the largest in the world (automotive manufacturers like Volkswagen and Toyota occupy top ten slots in global corporate revenues). Rapid transit is a priority for policymakers and individuals alike. In-car gadgets and cheap flights to the Med are wonderful to have. But when it comes to getting from A to B, we've been thinking small. We've become attuned to a steady but settled rate of progress, to the sense of speed, yes, but not acceleration. Again, we have a breakthrough problem.

*

This pattern resembles the fields of pharmaceuticals and medicine. In the early days, individuals make breakthrough progress. Over time those breakthroughs speed up, accumulate and cross-fertilise. At some point, although more and more investment goes in, the fundamental outcomes stop changing as much. Life expectancy improvements slow down or stall; average speeds remain constant. The days of the Wright Brothers, like those of Pasteur, are gone.

There are signs that things *could* change. After decades of incrementalism, we are starting to see the outlines of something new. Electric cars are becoming ubiquitous; the reign of the internal combustion engine is ending. Likewise an even more fundamental shift in the nature of transport is undergoing a colossal research effort: autonomy. Arguably, autonomous vehicles are already the most promising and advanced breakthrough, in that a fleet of them, working as one great hive mind, would herald a revolution in the transport system.

It may seem that supersonic travel is for the history books. Yet aeroplanes are subject to possible reinvention, with companies exploring speeds of up to Mach 5 – London to New York in ninety minutes. Delivery drones like Amazon's Prime Air programme are at advanced stages of testing; startups in Germany, the US, China and New Zealand are trialling flying quadcopter cars; solar-powered planes circumnavigate the Earth, and prototype jetpacks are available to buy if you have enough money. Space is reopening thanks to a generation of buccaneering entrepreneurs like Elon Musk, Richard Branson and Jeff Bezos. In addition to the old heavyweights of NASA, Russia and the European Space Agency, China and India have booming space programmes. The Chinese Tiangong programme will soon have fully fledged space stations in orbit. Ambitious lunar and Martian missions are planned for the near future. NASA's Artemis mission is optimistically mandated to go back to the Moon by 2024.

We can easily imagine a future of bustling skies filled with great airships, tiny darting drones, hypersonic intercontinental jetliners, solar-powered long-distance cruisers, blizzards of AI-controlled car-like transportation systems, adrenalin junkies getting to meetings on jetpacks, information teleported around the world in real time. We can easily imagine it, and, in fact, we have done for a century ... But despite that, it's not here. We are fairly good at the conception stage, less so at the execution and, especially, purchase phases. Many if not most of these technologies are still in a perilous untested state. To reach this point has already taken volumes of R&D capital massively in excess of those deployed for the creation of previous technologies. As we saw, it did not take teams of PhDs to build the Flyer.

We have never invested so much in cutting-edge modes of transport since the high days of the space race. Private and public capital is pouring into areas like space, drones and autonomous vehicles; the big incumbents like Boeing and Volkswagen are finally waking up to the challenge and putting serious efforts into radical change. The potential is clear; but it's not unreasonable to wonder why it's not translating.

Medicine and transport affect everyone. They are global priorities connecting scientific and engineering frontiers with politics and policy. In each field, breakthroughs made in the past are fundamental elements of modern life. But there is evidence of a misfiring; each has a breakthrough problem.[42] Assuming that we will automatically carry on coming up with big ideas quickly and easily is a mistake. The stagnation thesis has teeth. Every age has its breakthrough problems, but that's the point: despite our success we haven't bucked the issue.

What happens in future will require us to understand and confront this problem, just as people in the past managed to confront their own manifestations of it. It is the great question of our times, just as it was for them. Nor is it simply an unfounded nostalgia for an age of jets and spaceships that never was. A generation of economists has quantified this stalling. They show it isn't just an intangible feeling of wanting the impossible, a dim sense bred of reading too much science fiction. Rather it's grounded in the economic data. Medicine and transport are not alone; they're canaries in the coalmine.

3

The Diminishing Revolution

1873 and All That

Perhaps the most important scientific, technical or cultural achievement of the year 1873 was the publication of James Clerk Maxwell's *Treatise on Electricity and Magnetism*, the summation of a game-changing career in physics. In Einstein's words, 'One scientific epoch ended and another began with James Clerk Maxwell.'

Running to over 1000 pages, Maxwell's book incorporated the sum of knowledge about electricity and magnetism. At once proto-textbook, deep overview and conceptual pioneer, the *Treatise* notably introduced the idea that electromagnetic waves exert radiation pressure: the startling proposal that sunlight exerts around 4 lb of pressure on every square mile – something experimentally validated in 1900. Maxwell's biographer argues that the *Treatise* is 'an explorer's report', second only to Newton's *Principia* in importance as a physics or even a science book.[1]

Before Maxwell, electricity and magnetism were subjects of growing interest. Thinkers like Ørsted, Coulomb, Ampère and

Faraday, whom Maxwell would come to know, had shown that the two forces must be linked. But exactly how was still a mystery; known facts suggested fragments of a theory, but it pushed at the edges of contemporary science and maths. In the mid-nineteenth century, the puzzle remained. At the core of Maxwell's achievement lies his uncovering of the laws of electromagnetism, uniting the two and solving the mystery in a brilliant series of insights.

Born in 1831 in Scotland, a precocious child growing up comfortably amid the wide landscapes of Galloway, Maxwell was interested in the world from an early age. While at the Edinburgh Academy, aged just fourteen, he produced his first paper, on the mathematics of curves. Stints at the University of Edinburgh and Trinity College, Cambridge put him at the heart of Britain's changing scientific world. Before long he had posts as a professor of Natural Philosophy in Aberdeen and London. Aged just thirty he was elected a Fellow of the Royal Society. Maxwell's career came at a turning point for science.

Using radically imaginative models, Maxwell teased open the nature of electromagnetism, realising that light was part of the same puzzle. Previously obscure connections came together in a series of beautiful equations uniting the three extraordinary properties: 'one of the greatest scientific discoveries of all time'.[2] In time his work on electromagnetic waves led to innovations like radio, television and radar; it has been described as the second great unification of physics after Newton.

But Maxwell's importance is wider still. He made advances in the understanding of colour, vision and light, work that ultimately led to the world's first colour photograph, debuted at London's Royal Institution in 1861. Later his three-colour principle was vital for the invention of colour television. He made vaulting gains in the understanding of Saturn's rings, then one of the most intractable problems in planetary physics.

Before moving to electromagnetism, Maxwell had theorised the radical idea of a field. His understanding of gases led towards the use in science of statistical models, a mathematical advance that paved the way for modern physics. Maxwell is pivotal here: after him, physics grew ever more abstract, conceptually reliant on the most sophisticated mathematical techniques. Maxwell understood that while some processes were inaccessible to direct human perception, statistical virtuosity could bridge the gap. In the paper *A Dynamical Theory of the Electromagnetic Field*, a seven-parter introducing what are now known as Maxwell's equations, he displayed the full power of this new physics: an abstracted approach showing how the relationships between electricity, magnetism and light all flowed from the laws of dynamics. Not until twenty years later did Heinrich Hertz produce and detect electromagnetic waves, confirming Maxwell's theory. Again, Maxwell was ahead of his time; his theory had to wait for experimental validation, now the common pattern in physics.

His thought experiment – Maxwell's Demon – is one of the most influential in the history of science. Maxwell was also instrumental in standardising a system of units for the study of electricity and did early work on topology, while his research on machinery, control and feedback helped found the field of cybernetics in the 1940s. To top it off, he was the inaugural director of Cambridge University's Cavendish Laboratory. Although third choice, he was an inspired leader for the new laboratory, which had been expressly created to catch up with a professional era of dedicated labs. In time the Cavendish would host dramatic breakthroughs like the discovery of the electron and the structure of DNA.

When Maxwell died of abdominal cancer in 1879, science had been transformed. His insights into electricity, light and magnetism are among the greatest intellectual leaps ever accomplished. If we zoom in on 1873, on Maxwell's *Treatise*, we find the genesis of modern physics, engineering and institutional science. Maxwell is a hinge, a catalyst.

As with his contemporary, Pasteur, it's difficult to imagine any individual or team, perhaps even any institution, having a proportionate impact today: not just one big idea, but reams of them, again and again entirely reorienting disciplines and worldviews. Even giants of mid-to-late twentieth century physics, field-shapers like Richard Feynman or Murray Gell-Mann, couldn't have such universal impact. Will it ever again be possible for one researcher to alter the nature and trajectory of their field so profoundly, as Darwin, Pasteur, and later Freud, Poincaré, Turing or Einstein did? So far it looks like yet another example of the breakthrough problem. But there is more to the year 1873 than the publication of Maxwell's *Treatise*.

In 2005 a hitherto obscure physicist working at the Pentagon's Naval Air War Center in California made a stir with a paper entitled 'A possible declining trend for worldwide innovation'.[3] Jonathan Huebner wanted to puncture triumphalist narratives of scientific and technological success; he didn't believe they followed an ever-escalating curve. Instead he claimed to detect a reversal. The curve of innovation, he argued, at first rose slowly over millennia, then massively increased before starting to tail off at the moment of its greatest triumph.

The data suggested a peak year for human innovation, and it wasn't in the late twentieth or early twenty-first centuries. It was, he said, 1873. From the whole of history Huebner had landed on this arrestingly specific and un-anticipated date.

Huebner started by looking at the rate of innovation, defined as the number of important technological developments per year divided by world population. This gives you a rough measure of how much each individual contributes to a given technological development. Analysing 7198 scientific and technological developments from the Dark Ages to the present, he plotted them against estimates of world population drawn from the US Census Bureau.[4] Doing this suggests that the rate of major innovations per head peaked in 1873; after that date the development of a major scientific or technical innovation required ever more people.

Population is a blunt instrument, but the picture looks worse if you examine the ratio of innovation against either GDP or education expenditures, which are arguably better indicators. Per capita GDP grew by a factor of 9.62, whereas population only increased by a factor of 3.77. The result is that 'the number of innovations when normalized to world per capita GDP declined 2.55 times more rapidly during the twentieth century than when normalized to population.'[5]

Education expenditures, recorded as a proportion of GDP, went up by even more. Adjusting for investment in education, the decline is yet steeper. Many of the innovations listed from the late twentieth century, moreover, are only improvements of earlier technologies. Looking at patent numbers against population shows a similar pattern, although with a later skew: the peak here is 1916. Better to have been born in the nineteenth century than the twentieth or twenty-first if you want to create a great invention or make a dazzling discovery.

His paper caused immediate controversy. Some picked up on the story as newsworthy naysaying in a time of apparently accelerating change. The year 1873 was eye-catchingly specific and counterintuitive. The most telling criticisms centred around the list of innovations themselves. Wasn't it factitious to make such broad claims on one contentious reading of history? His list may be well-considered, but it is inevitably somewhat arbitrary. The paper relies on the equivalence and validity of the selected innovations, discoveries and technologies, and this is a matter of debate to say the least. Critics like Ray Kurzweil also believed that it missed how important current innovations could become. Others thought population a poor measure.

But Huebner's argument is none the less revealing. Some years later, the geneticist Jan Vijg looked at the number of significant inventions per decade as judged by Wikipedia's timeline of historic inventions.[6] Vijg argues that Wikipedia's self-correcting, deliberative model makes it particularly valuable here. It showed a similar pattern: a rise from the 1830s to the end of the century, a genuine golden age in the early twentieth century and a later drop-off. The 2000s were a notable low.[7]

Huebner's paper is, if not conclusive, then suggestive. It was an early shot in the debate, posing an uneasy question about the arrow of innovation. It implied that the breakthrough problem isn't isolated in a few significant areas, but operates across human civilisation; that big ideas are subject to diminishing returns.

And it focuses attention on a fascinating period around 1873, the time when Maxwell transformed physics and Pasteur invented modern medicine, the era of our near ancestors' life transformation. 1873 isn't a year plucked at random. Maxwell and Pasteur worked at a particularly fertile fulcrum in the timeline of big ideas. Widen our view just a little, and 1873 does indeed have the potential to be the pivot of history.

Industrial revolutions are clusters of big ideas: in science, technology, economics; in how we harness energy and collect, store and disseminate information; in organisational form and business models; in how we live, what it means to work and play, what can be produced and consumed. They are Big Big Ideas, a paradigm of paradigms all shifting at once. 1873 stood on the cusp of what came to be known as the Second Industrial Revolution (2IR). If the First Industrial Revolution

(1IR) lit the touchpaper of modernity, the 2IR saw it catch fire. In the words of the polymath researcher Vaclav Smil, who places this birth of a new world in the years 1867 to 1916, it 'created the twentieth century'.[8]

We have already encountered the 2IR in the Introduction. It was here that the conveniences of modernity came on stream, from electric light to clean running water, elevators to large multinational corporations, production lines to the beginnings of social security, consumer brands to telephones, radio to the tabloid newspaper, moving pictures to motor cars. If Maxwell represents the birth of a new physics, a brief survey finds an equivalent innovator or innovation in almost any field of endeavour.

In the UK, the 1IR's heartland, the first wave of industrialisation started to level off from 1825.[9] In the background, important research continued and steady advances in areas like electricity, metallurgy and chemical engineering later proved decisive. The 2IR, like the first, like all major breakthroughs, represented decades of work. Soon this was evident in a set of widely used innovations. Founded on a new material and energy base, they were not just big ideas in themselves, but the building blocks of other ideas.

In material terms this was an age of steel, newly abundant thanks to the Bessemer and later the Siemens-Martin process. In just a decade, 1867–1877, annual American steel production grew from 20,000 tons to one million, with prices falling by two-thirds.[10] The 1IR had been powered by coal; the 2IR hit upon a new, even richer energy source: oil. In the twenty years between 1859 and 1879, American oil production grew from 8500 barrels of refined crude to 26 million. In parallel the price fell from $16 a barrel to just a dollar.[11] The two technologies had a host of knock-on effects: buildings could be taller, and transport more efficient and powerful.

Macro-inventions came fast. New chemicals from dyes to dynamite, aspirin to fertiliser were harnessed and mass produced. Electricity generation and the electricity network, a general purpose technology surely as significant as any in history, came of age; then advances were made in the manufacturing process itself, a total system exemplified by Fordism with its interchangeable parts and production line. Changes in food supply from the use of nitrates to refrigeration improved quality and length of life; and a revolution in communications media introduced a now familiar informational mix. In the social and political sphere explosive ideas of political equity, democracy and revolution arose; in the aesthetic sphere there was radical experimentation; and in business came new conceptions of private enterprise and the organisation of work.

But the creation of a recognisable modernity was a wider endeavour. If you look aside from epochal inventions like the telephone and the internal combustion engine, or the rollout of electricity, specific decades like the 1880s still loom large. Those years alone saw such innovations as revolving doors, skyscrapers and lifts, ballpoint pens, Coca-Cola, vending machines, cash registers, electric railways, irons, bicycles and antiperspirants, among many others.[12] Joel Mokyr argues that the technical changes were so important because, for the first time, the deep principles behind why things worked were understood and could be utilised.[13]

In terms of control over our environment, our ability to master and manipulate it, no other moment in history rivals that of the pre-First World War world.[14] If the 1IR was a break with the general pattern of history, the pace of the 2IR's macro-invention is unmatched, setting heightened expectations for delivery of ideas. Huebner has a point: 1873 is no arbitrary date, but the dawn of an explosive pulse of big ideas.

We too are living through an industrial revolution, the third

(3IR, roughly since 1970). The 3IR is all about digitisation, the equivalent of steam or electricity. Jobs, consumption, knowledge, social lives and relationships seem to have been swallowed whole by the 3IR. And now commentators posit the existence of a fourth (4IR). Nonetheless the Stagnation Debate rages about the relative value and scale of this new industrial revolution.

In the world's wealthiest countries, those closest to the frontier, long-term economic growth as measured by GDP is in a pronounced pattern of slowdown.[15] The growth rate in real per capita GDP (in the US) has slowed from 2.25 per cent, smoothed out across the twentieth century, to 1 per cent in the twenty-first century.[16] If anything this slowdown in growth is becoming more marked as the century goes on. This seems strange, to say the least, in a time when tech and economic understanding should be spurring growth on. Productivity figures, a useful shorthand for the delivery of new ideas within an economy, are also significant. Productivity growth has been much slower over the 3IR than before. Since 1970, Total Factor Productivity (TFP), the key measure of how technology boosts growth, has grown at only a third of the pace achieved between 1920 and 1970, leaving us fully 73 per cent behind the postwar trend.[17] In the words of Tyler Cowen and Ben Southwood: 'TFP growth probably is the best contender for how to measure scientific progress. And overall TFP measures do show declines in the rate of innovativeness, expressed as a percentage of GDP.'[18]

In short, it appears that recent ideas have failed to have the same impact as those of the earlier generation. Even the US Congressional Budget Office and Federal Reserve have begun working with forecasts of lower long-term productivity growth.[19] We have gone from making major gains to marginal improvements, and the results are commensurate. The prognosis for big ideas is on shaky ground.

For many this was the smoking gun in the Great Stagnation Debate, at the very least prompting questions about the scale of innovation at work. This is still a matter of a debate among economists; some see stagnation in science or innovation, others view the impact of demographics and a shift towards the slower growing service sector as primary drivers.[20] Other measurements, however, support the picture of a slowdown in innovation. Patents have increased in absolute numbers, but not in quality. Recall that patents are one area where researchers have established quantitative models for novelty and impact; those same researchers conclude that the quality of patents, judged by how original and influential they are, was higher in the 1850s and 1860s than it is today.[21] In essence patents may be showing a lack of improvement in the generation of big ideas since before the 2IR, while nearer to our time, the numbers of such breakthrough patents have noticeably dropped since the mid-1990s, just as the 3IR really hit its stride. Overall, whatever the cause of the growth slowdown, it's noteworthy that technological innovation is not overcoming it. Economic growth and impact at the frontier seems harder to come by today, and this strongly supports the stagnation thesis.

During the 2IR, change happened across every dimension of human experience, from housing to communications, transport to healthcare. In contrast, argues economist Robert Gordon, the 3IR concentrated fundamental advances – big ideas – in entertainment, information and communications technologies. Yes, these are significant. But Gordon likes to ask people: if you could have all the innovations from the mid-1990s on, or access to hot and cold running water in your house, which would you choose? Most of us probably wouldn't take the joys of social media over not having to haul gallons of water on cold winter mornings, or having electric light in the evening, or a car. Aside from informational and communications goods, the signature advances of modern life occurred decades ago: not only light or the car, but electricity, refrigeration, transport infrastructure, household labour-saving goods.

Whether measured in output per hour or per person, not only has the 3IR seen slower growth, but also 'the unmeasured improvements in the quality of everyday life created by 3IR are less significant than the more profound set of unmeasured benefits of the earlier industrial revolution.'[22] In the words of the economist Martin Wolf, 'We're living in an age ... of really slow and boring technological change compared to what our ancestors managed to generate.'[23] Or Peter Thiel, again succinctly expressing the technological stagnation view: 'I don't think we're living in an incredibly fast technological age.'[24]

In reality, the 3IR has funnelled progress into the area of least resistance: software.[25] Whereas the 2IR saw innovation across almost every endeavour, the 3IR boiled it all down to little screens. It allows work in a frictionless and easily malleable sphere, the contained and supple world of code. Grad students and dropouts could build the world's major software corporations. One or two graduate students could not deliver a redesigned jet engine or nuclear power plant; barriers in the physical world are too high. Whereas the 2IR built the modern world, the gains and innovations of 3IR are narrower, its economic impact surprisingly muted at the macro level, debatable according to the World Bank.[26] Digital technology gave the sensation of acceleration, but exponential improvements like Moore's Law cannot be ported over to the material world.

Many of the 'unprecedented' marvels of the 3IR have deep roots: a century before Amazon became a technologically powered 'everything store', Sears Roebuck's mail order business fulfilled 100,000 orders per day from a catalogue of almost

1200 pages. They even built the world's largest commercial building, complete with miles of mini-railway for processing orders. We think of Apple and Google as sizeable, powerful and innovative beyond parallel, but forget that AT&T once dominated the Dow Jones index, employed a million people, held a monopoly on US telecommunications and invented the foundations of modern computing (more of which later). The Internet didn't destroy distance; it was already dying, mortally wounded since the invention of the telegraph in the 1840s. We have graphene; the 1930s had nylon, neoprene and Teflon. We have genetic engineering, but its foundations were laid over a century ago.

We often think of our time as one of unique acceleration.[27] But this is an illusion. We just haven't adjusted to a new reality: slowdown. For example, per capita energy use grew for centuries in what has been called the Henry Adams curve, underpinning technological and societal change.[28] This growth has stopped: if pre-1970 energy use trends had continued, we would have access to thirty times as much energy today, likely supplied by novel forms of nuclear power.[29] But at the frontier, the availability of energy has stagnated, even as it is growth in available energy that underwrote almost everything about modernity.[30]

Across a range of significant metrics there is a marked deceleration in the rate of change: especially in developed countries, we are seeing a slowdown not in just absolute economic growth, or major innovations, or energy use, but also in debt (student, automobile and mortgage), the number of books published per year, population and, with it, fertility rates, the number of relationships and age of marriage, improvements in living standards, median wage growth, property and other asset price inflation, increases in human height, the introduction of new significant consumer appliances and enrolment in tertiary education.[31] Covid-19 has witnessed dramatic slowdowns in business, hospitality, transport, the number of conferences and the pace of life. This slowing – and in some cases outright reversal of growth – isn't necessarily all bad but it does imply we have hit an historic inflection point.

In contrast to the past, the breakthrough problem spans fields, its tentacles suffocating the economy. Gen Z – my children's generation – will have lives broadly similar to my own or, worryingly, worse. Their houses, jobs, health, appliances, work, media consumption and lifestyle will be quite recognisable. That was not true of my parents and their parents, or even more my grandparents and great-grandparents. Beyond the screen, 'the frontier of last resort', this is an age of consolidation.[32]

And what of a putative 4IR? We will return to this later, but it's worth remembering that the 4IR is as much a marketing gimmick of the World Economic Forum, which coined the term, as it is a coherent concept. The 4IR is less a historical fact, more a loosely assembled set of conference talking points notionally about 'cyber-physical systems', a handy label for a basket of potentially transformative but still nascent technologies that face a host of technical, ethical and social barriers.

In the meantime innovation has been directed at ways to order pizza or take better selfies; the softest frontiers around. To be fair, Google Maps, Zoom, Minecraft, Spotify – these are marvels and big ideas that have found purchase, and digital is the bright spot. But it only makes the contrast with other areas starker, and doesn't fully explain why measures like TFP have fallen so much. While we shouldn't play down the significance of the 3IR, its impact doesn't equal let alone exceed the 2IR.[33] It adds further context to Huebner's 1873 hypothesis and its many heirs: that far from living in an escalation of big ideas, we live in a confused, halting sort of advance.

This is the crux of the case for the Great Stagnation. Our record of delivering breakthroughs seems to be getting worse, our revolutions themselves diminishing. Our ideas appear to matter less at the frontier.

What price a breakthrough?

For centuries economics was plagued by a contradiction. On the one hand, economic thought suggested that competition in free markets eliminated competitive advantage; eventually everyone reaches the same point and diminishing returns set in. On the other hand, there was evidence that efficiencies produced by increasing scale could, against the idea of open competition, maintain growth. Both strands go back to Adam Smith's foundational work: on the one hand his pin factory, which, as it grows and creates new specialisations, produces ever more pins. And on the other the invisible hand, working its magic, leading markets towards the nirvana of equilibrium.

While on the ground there seemed to be evidence that returns were increasing, in that big companies tended to get bigger, the mathematics of markets indicated that they should tend towards an equilibrium where profit evaporated in the face of sheer competitive force. The mystery also extended to answering a further fundamental question in economics: how does economic growth occur? As David Warsh records in his magnificent book, *Knowledge and the Wealth of Nations*, it wasn't until the 1980s that the puzzle was solved.[34]

This is the story of how ideas became central to economics. Understanding how ideas are formed and shared, where they come from, how they work, and crucially, the mechanisms, tools, incentives and costs in their creation, went from peripheral concern to the heartland of modern growth theory. Instead of the traditional economic inputs of labour, capital and land came a vision of the economy and economic growth as 'ideas, people and things'.[35]

Until the late twentieth century, the central account of growth came from MIT economist Robert Solow. Developed in the 1950s, his understanding, 'the workhorse of economic growth models', became the dominant postwar paradigm.[36] Solow looked at how the inputs of labour and capital drive growth.[37] Add more of them, deliver growth. Crucially, technology made them more productive: it was the 'residual', the essential element unexplained by labour and capital alone. The residual might include temporary boosts in government spending, but was predominantly technological in character.

Solow convincingly showed how times of fast technological change delivered economic growth. But this technology wasn't part of the system as such; it was 'exogenous', an external, largely unexplained and autonomous process. And yet the sums showed it accounted for the vast majority of growth. Increases in output explained by technology were known as Total Factor Productivity, the same measure that has been waning in our own age. Clearly there was room for further elaboration.

In the 1980s a young economist called Paul Romer made decisive advances on Solow, offering a new understanding of growth and helping solve a great riddle of economics.[38] Romer's insight was that ideas were at the heart of economic growth. Whereas Solow bracketed technological progress outside the economic system, Romer elucidated a model where technology was 'endogenous', within the economy. Researchers and entrepreneurs weren't outside the system, but were caught up in its incentives and mechanisms. Technology wasn't some airlocked programme beyond economics; instead it was driven by it.

As outlined in a now famous paper – 'Endogenous

Technological Change' – the significance of ideas became apparent. For Romer, ideas lay behind the wonders of the modern world. Growth was their by-product. His analysis of economic history made the case that, wherever you found new, wealth-generating technologies or goods, you found new ideas, new ways of combining things; sets of instructions, as he saw them: 'A hundred years ago, all we could do to get visual stimulation from iron oxide was to make it into pigment and spread it on fibers that are woven into canvas. (Canvas itself was a big improvement over cave walls.) Now we know how to spread iron oxide on long reels of plastic tape, and use it with copper, silicon, petroleum, iron, and other assorted raw materials that have been mixed together to make television sets and video tape recorders.'[39]

Knowledge and ideas are central ingredients in a flourishing economy because they have different properties to other economic phenomena. Economics traditionally studied the allocation of scarce resources; labour, capital, land, material things. All were in finite supply. In technical terms they were rivalrous: if I have an apple and eat it, you cannot eat the same apple. Ideas by contrast are non-rivalrous, undiminished by sharing or consumption. Inherent in this picture was the notion that ideas could, by being non-rivalrous, allow for increasing returns. In other words, ideas prompted a new kind of economics.

Now incorporate another of Romer's key insights: ideas aren't just non-rivalrous, in the modern economy they are also 'partially excludable'. By a combination of legal instruments like intellectual property (IP) and commercial secrecy, ideas can, for a time, be protected. Some people can have theoretically exclusive access. But in the long term they spread, go out of patent, spark copycats and so add to the public stock of knowledge. The mechanisms of partial excludability explained how increasing returns were possible in open and competitive markets. They explained how researchers and innovators were incentivised, how the whole process was brought inside economics, no longer a residual beyond explanation. They placed ideas centre stage. Society's institutions for discovering, nurturing, protecting and, ultimately, disseminating ideas were no longer the province of philosophers and historians but moved to the bleeding edge of economics, helping enable the reappraisal of ideas' historical role we saw in Chapter 1. This focus on ideas is borne out in the economy: while real US GDP doubled over the twenty years after 1980, the physical weight of its non-fuel inputs remained essentially constant.[40]

Romer's revolutionary paper, a 'watershed',[41] is itself a fantastic example of a breakthrough idea, one that eventually won him the Nobel Prize in Economics. But it left some questions unanswered. Romer assumed that if you invested in new ideas at a constantly growing rate, an equivalently constant rate of return would be exhibited. If you have more researchers and their productivity improves, this will automatically lead to more economic growth. Boosting the number of researchers or R&D spend should, therefore, boost economic growth. Put more into big ideas, and you should get more big ideas.

In the mid-1990s a young Stanford economist, Charles I. Jones, began questioning that assumption. Just as Romer had put the study of ideas on a firm economic footing in his 1990 paper, so five years later Jones introduced a wrinkle.

Jones noticed something interesting about growth theory: the output of ideas wasn't commensurate with R&D.[42] Between 1950 and 1987 the number of US-based scientists and engineers had risen from under 200,000 to a million, a fivefold increase not reflected (even remotely) in the growth rate.[43] Aggregate R&D went up, average growth rates did not. Jones had found evidence that ideas were, in a material sense, getting harder to realise.[44] Again, it seemed to suggest, ideas were subject to diminishing returns. What did that imply about the future of the economy, of growth, of humanity, of ideas themselves? Jones's later work suggested a further disturbing trend: that much of the growth in the latter half of the twentieth century was based not on new discoveries, but much older ideas.[45]

A more recent paper, 'Are ideas getting harder to find?', from Jones, along with colleagues Nick Bloom, John Van Reenen and Michael Webb, builds on this work.[46] Here detailed empirical research develops an extraordinary new model giving concrete shape to the struggle for new ideas and showing how, over time, their generation has become more difficult. At its heart are Romer's insights that economic growth relies on the creation of new ideas, and the creation of new ideas relies on researchers and their productivity. But while the number of researchers and R&D expenditure is going up, their research productivity is going down. It takes more and more research to produce new modes of transport or new life-saving drugs. As Huebner had clocked, our ability to generate new ideas seems to diminish with every passing year.

A sample of firms reveals that research productivity is declining in 85 per cent of the companies surveyed. On average, research productivity was decreasing at 9 per cent per year: compounded, a huge fall.[47] The authors also control to ensure that it isn't simply that new ideas are coming in, powering growth and building new sectors, while older ideas get more and more stuck. Unsurprisingly this translates to the wider economy.

They look at trend growth in the US economy over long periods and find stable or declining growth. But, on the other side of the ledger, research efforts have hugely expanded. If we were constantly finding fresh turf for quick gains, opening up unexplored areas of discovery and business, these new ideas and sectors should show up. Since the 1930s, research effort has increased by a factor of twenty-three, averaging growth of 4.3 per cent per year. But total US research productivity has gone down by a factor of forty-one since the 1930s, a decline of 5.1 per cent a year. Another way of thinking about this is to say that every thirteen years we need to double our research effort just to stay on the same course.

The trend is stark. It's not just about what it represents for economic growth, wellbeing and the dynamism of companies. It has potent implications for the very structure of new ideas in the world.

Bloom et al study the critical foundation of the 3IR and the digital economy: Moore's Law, that opposite of Eroom, named after the co-founder of Intel, Gordon Moore, and predicting that the number of transistors on an integrated circuit, a chip, doubles every two years. However, as the paper argues, the actual story is more complex. While Moore's Law looks like a rising technological curve, maintaining that curve requires more and more effort and expense. Research productivity in computer chips is declining at an average of 6.8 per cent a year: 'Put differently, because of declining research productivity, it is around 18 times harder today to generate the exponential growth behind Moore's law than it was in 1971.'[48]

Crop yields and agricultural R&D exhibit a similar pattern. Since the 1980s growth in agricultural production has been slowing, but the number of US researchers in the area doubled between 1970 and 2007. For crops including corn, soybeans, cotton and wheat, yields are generally on long-term rising curves, doubling between 1960 and 2015. But R&D expenditure from both government and non-government sources, including research on cross-breeding and hybridisation for better insect resistance and nutrient uptake, improved herbicides and pesticides, bioengineering and the automation of seed-related tasks, has also risen sharply. Just as with Moore's Law, it takes more and more researchers to maintain the same level of growth. Depending on the crop and the R&D effort, the increase factor ranges between 3 and 25. The average productivity growth in corn yields equates to minus 9.9 per cent over the period.

Jones, Bloom, Van Reenen and Webb also pick up on the healthcare question. They examine 'new molecular entities' (NMEs) approved by the Food and Drug Administration. These may be chemical or biological, and they include virtually all new and significant drugs. Between 1970 and 2015 research inputs increased nine times, but over the same period research productivity fell by a factor of five. In other words, the number of new NMEs has risen, but it takes a vastly greater number of researchers to develop each NME now than in 1970. Eroom strikes again.

But they also wanted to look at how diseases are treated more widely, and the relationship between this, idea production and R&D. Life expectancy rises in a linear, non-exponential fashion, but such arithmetical increases are correlated to exponential growth in research. The result is, once again, a decline in the efficiency of our efforts. The paper examines the two top killers in the US: heart disease and cancer. Using publications relating to a given disease as a proxy for research efforts, they find that the number of publications on cancer increased 3.5 times in the years 1975 to 2006, and publications on clinical trials 14.1 times. Despite these significant increases in research effort, the additional years of life saved is falling: 'between 1985 and 2006, declining research productivity means that the number of years of life saved per 100,000 people in the population by each publication of a clinical trial related to cancer declined from more than 8 years to just over one year. For breast cancer, the changes are even starker: from around 16 years per clinical trial in the mid 1980s to less than one year in 2006.'

The paper was quickly picked up by the media, discussed at conferences, on blogs, in op-ed pieces.[49] Here was seemingly hard proof that in a wealth of fields, the creation of ideas is getting harder. If you judge research output by verbiage, you don't find a problem; if you judge it by the rate at which ideas find purchase, you do. The same analysis could be given regarding transport, for example: it would likely exhibit a strongly rising curve of R&D spend, not matched by increases in travel speed or comfort.

If Romer's work on endogenous growth theory was vital in bringing ideas into economics, Jones and his colleagues show how ideas are not as ceaselessly abundant in new varieties as we might expect. As the authors say, 'research productivity is falling everywhere we look.'[50] It seems the frontier has become sticky. This indicates a deeper trend undergirding the breakthrough problem. It adds heft to the pattern of diminishing returns noticed by Huebner and provides a mechanism behind the stagnation thesis.

Yes, this work doesn't distinguish between what I call breakthrough ideas and any old idea. Romer's concept of ideas is very different to the wider sense of the term I am using. But it provides stark evidence that a fraught environment now exists for big thinking. And Jones and his colleagues aren't alone among economists. Evidence is mounting from across the field.

*

Economists are increasingly sceptical about claims that we live in an age of radical innovation. High-profile names like Lawrence Summers, Robert Gordon and Tyler Cowen elaborate the idea of 'secular stagnation', citing that strange slowing of Western economies despite a surface-level technological abundance. Cowen's 'Great Stagnation' is exactly what you'd expect if research productivity were to decline at 5.1 per cent a year.[51] Prior to the 1970s, living standards doubled every couple of decades. Median wages have since been roughly flat. Median income doubled between 1947 and 1973, but in real terms grew only 22 per cent in the period 1973 to 2004.[52] If the postwar growth rate had continued, by 2010 the median US family income would have been $90,000. Instead it was $54,000. Permanent low or even negative interest rates hint that we don't have sufficient good ideas that compete aggressively for funding.[53]

New technology should trigger the creation of new businesses. The invention of the motor car, for example, produced jobs in everything from spark plug factories to car park construction. Big ideas create big new sectors, which open space for new companies. Hence the rate of big idea creation should result in rapid business formation and overall economic dynamism. Alas, the economy does not paint an encouraging picture of an era brimming with big ideas. Even pre-Covid it was gasping for new thinking. Company creation is at its lowest for thirty years or more. Only a handful of companies in Germany's DAX and the French CAC were founded after 1970.[54] Of Europe's most valuable 100 firms, almost none were created in the last forty years.

Although America does better, the pace of successful business formation there is still poor. Startups struggle. Their number is falling and the failure rate is abysmal: 95 per cent fail to deliver the expected returns and 30–40 per cent burn through all their capital.[55] Most surprisingly of all, the numbers of new firms and IPOs (Initial Public Offerings, i.e. stock market launches) both peaked decades ago.[56] Rates of entrepreneurship have declined in major economies like the US, the UK and Germany, and they have declined faster among people with higher degrees.[57]

Another canard is that we live in times where awesome new technologies are toppling industries at the click of a distant button. Again, prior to the pandemic, data suggested that rates of such job destruction, a sure indicator of the invention and rollout of impactful technology or practices, were also falling. Not only are there fewer new businesses; there are even arguably, against the popular perception, fewer so-called 'unicorns'– startups with a billion dollars or more capitalisation.[58] And the pattern holds even in the fabled tech sector: tech company creation peaked in 2000 and growth rates have been on the slide since the 1980s.[59] Silicon Valley is dominated by walled giants often more interested in their survival and profit than radicality.[60] Indeed, their sky-high market capitalisations may reflect a growing awareness of permanent lock-in at the top. It all suggests that business ideas are thin and innovation overhyped. Far from the past few decades representing a Great Disruption, they are closer to stagnation: job destruction has slowed down even faster since the year 2000.[61]

Most advanced economies display a clear oligopolistic tendency: in banking, energy, telecoms, property, consumer goods and food retail, for example, the major players have remained essentially static for decades.[62] In 1987 only one-third of firms were more than eleven years old; by the 2010s it was half. Moreover, these old firms account for 80 per cent of the total US workforce, up from 65 per cent in 1987.[63] The proportion of young firms correspondingly continues to fall. Another of way of looking at this is that in the 1970s, about seventeen establishments opened for every 100 existing ones, while thirteen closed. After 2000 the corresponding figures have been more like thirteen and eleven – fewer entrants, fewer exits, less dynamism overall.[64] The US government estimates that Fortune 500 companies doubled their share of the economy between 1955 and the present. Even here the top take more: the share of Fortune 500 revenue going to the top 100 grew from 57 per cent to 63 per cent between 1994 and 2013. The share of top patents going to established older firms has risen steeply, while new companies increasingly struggle to reach the frontier.[65] Over roughly the same period the number of publicly listed companies halved thanks to consolidation.[66]

Moore's Law may require more and more input to keep going, but what of fields still in their freshest, most burstingly productive phase? Capturing such information isn't easy but Huebner and Vijg provide two negative examples, and the authors of the ideas paper do attempt to account for this in part. Sources like *Encyclopædia Britannica* and Wikipedia which both, in different ways, classify achievement also corroborate the general pattern.

Further evidence comes from a second look at patents. Major new inventions prompt not just individual patents, but whole new classes of patents. These in turn necessitate the reclassification of existing patents as those novel classes ripple through the firmament of invention. Big ideas should show up in either increasing or decreasing rates of patent class creation and class reclassification. While there is some ambiguity in the data, the rate (at least for US patents) seems fairly constant over time.[67] This buttresses the point about declining research productivity: despite having more researchers, knowledge and resource, we are not creating new patent classes any faster. If anything there is a slight bias towards a slowdown. Where the ideas paper looked only at existing technologies, whole new technologies are registered here; and here too we find diminishing returns.

*

A striking range of the world's leading economists sketch a worrying ground-state for the future of big ideas. They argue that the breakthrough problem is general, that diminishing returns are evident not just in medicine and transport but across swathes of the frontier. Whole societies appear to be running out of new ideas. When a tech titan like Peter Thiel or a science fiction writer like Neal Stephenson says they are disappointed with our achievements, it's easy to shrug: of course they'd say that! Platitudes about flying cars and settlements on the Moon are their bread and butter. But all this is more than anecdata. Research productivity isn't some distant measure; it ultimately affects every aspect of our lives, from how we communicate to what medicines are available at what price. The structure and rate at which it generates ideas is the measure of a civilisation; the elixir of success on the grandest scales and the crucible of small daily comforts. Anyone looking at the future needs to reckon with diminishing returns on the production of big ideas. Despite our successes and capabilities, no one can blithely assume a glorious neo-Enlightenment in the decades to come.

Economic arguments clearly play a role. But economists are (usually) the first to acknowledge they don't offer a complete description of the world. Big ideas are not just economic or technological. So what of frontiers beyond economically obvious innovation?

4

The Art and Science of Everything

1913 and All That

29 May 1913, at the Théâtre des Champs-Elysées in Paris
That evening, the impresario Sergei Diaghilev, his Ballets Russes company, the dancer Vaslav Nijinsky and the composer Igor Stravinsky created an artistic scandal. It revolutionised music and performance. The premiere of Le sacre du printemps – *The Rite of Spring* – came on a hot night on the cusp of the First World War. Nothing would be the same again.

The evening is fixed in the mythos of the twentieth century, one of the most debated, celebrated and studied moments in cultural and musical history. It was above all new, both aurally and in performance. As Nijinsky had written to Stravinsky in rehearsal: 'For some it will open new horizons flooded with different rays of sun. People will see new and different colours and different lines. All different, beautiful and new.'[1]

Born near St Petersburg in 1882 into a musical family, Stravinsky was plucked by Diaghilev from relative obscurity. Commissioned to write the ballet *The Firebird*, Stravinsky became an overnight Parisian celebrity. Here was a daring work breaking with the traditional sound of the ballet – embryonic hints of a paradigm shift in progress. Diaghilev had made a success of bringing 'Asiatic' Russian culture to a Western metropolitan audience and Nijinsky was already famous for a virtuosic gravity-defying style. Next the Ballets Russes and Diaghilev wanted to create a *Gesamtkunstwerk*, a total work of art, uniting music, choreography and stage design in one blistering vision. The ingredients were in place for something radical.

In the aftermath of *The Firebird* there was an appetite for more, and Stravinsky began speaking to designer and archaeologist Nicholas Roerich about ideas for the music, set and costumes. This was the genesis of the *Rite*, then called *The Great Sacrifice*, a work rooted in pagan Russian folklore and drawing on new theories of the theatre, heading towards the ritualised and non-naturalistic. But Stravinsky then got distracted writing his ballet *Petrushka* and the *Rite*, scheduled for 1912, was delayed.

Nonetheless Roerich and Stravinsky worked closely, initially at an artists' colony near Smolensk. Here came the idea of a ceremonial dance to the death. In October 1911, Stravinsky composed with startling intensity on the shores of Lake Geneva. With occasional previews before Diaghilev and luminaries like Claude Debussy, writing was largely finished by the end of 1912. Nijinsky began choreography. Rehearsals, on tour, amid squabbling, weren't easy, nor was the 'primitive', robotic and anti-classical dance. The orchestra was challenged by unexpected dissonances and raw noise. But everyone was aware that this strange, rhythmical work with its mutating metres and outlandish costumes was special. Final rehearsals took place in London (Stravinsky estimated it took 130 rehearsals, others more). The *Rite* was nearly ready.

Then came 29 May. Diaghilev had stoked the anticipation.

The well-heeled came out in force, but so did the artists: Jean Cocteau, Maurice Ravel, Claude Debussy. Picasso and Gertrude Stein were rumoured to be among the crowd.

In the hot theatre the programme started with more traditional fare. Then came the *Rite*. A note from a bassoon spiralled up from the orchestra while the curtains remained closed. This peculiar work quickly built into a complex collage of wind instruments. Trouble started just two minutes in. Boos, catcalls and hisses broke out in the auditorium. One contemporary witness described the coming of a 'tempest'.[2]

Things did not improve when the dancing began. It was heavy and deliberately ugly. The dissonance of the music, its tempo and violence all increased. So did the backlash. This wasn't music! This wasn't ballet! In Stravinsky's words, the storm broke: those who hated what they saw, which broke every rule, took their protest to the next level. Modernists vociferously countered. The music was drowned out by the tumult. The theatre was in uproar. People scuffled and spat.

Through it all the orchestra played on. Stravinsky, in the audience, anxious, eventually left to go backstage. As the first act went on, the protests grew. Some claimed that, as Diaghilev turned the house lights on and off, police were called to restore order and carted off several of the loudest protesters. The audience left knowing they'd seen something culturally momentous.

Critics were divided. Many hated it. But many others, even as they hated it, recognised the creation of 'a new music'.[3] Heavily dissonant, it was unlike anything people had heard, rewriting the rules of harmony, melody and rhythm. Repetition was deployed to shocking effect: primal, staccato rhythms with unusual metres. It was aggressively loud. At times the entire orchestra hammered away like some giant drum. Strings were shifted to the periphery, and wind, brass and percussion placed centre stage in sonically original forms. It was both ancient and vitally modern, machine-age but mythic. Scene breaks were stark, fragmented, almost Cubist. The whole thing was deeply disturbing. In the words of one critic, it was 'A new form of choreography and music. An entirely new vision, something never seen before, something gripping and convincing, has suddenly come into existence.'[4]

The long-term musical impact was huge. It pushed boundaries, becoming emblematic of the whole Modernist moment. Rhythm, the subject matter of ballet, the nature of noise, the scope for experimentation, the possibility of performance – all were changed.

Today *The Rite of Spring* is a classic example of a big new artistic idea. Its influence has no straightforward statistical measure; it cannot be assessed in album or ticket sales. Its contribution to GDP is negligible, its impact on human wellbeing unclear. But it is the quintessence of a big idea, a benchmark, purposefully ambitious and radical. And it was also part of a wider movement, where radical innovation in music and arts more widely became the norm. *The Rite of Spring* doesn't stand alone.

In comparison to the year that followed, it's easy to overlook 1913. Yet it repays examination. The world of 1913 was alive with fresh thought and uncoiling creative energy. There was a self-conscious effort to produce completely original ideas, to challenge everything that had gone before, a kind of big ideas project across creative and intellectual spheres. Few today would have the gall and gumption to announce their goal as the overthrow of art itself.

Besides the *Rite*, what else was happening? This was the year the City of Light went electric and withdrew the horse-drawn cart; the year it fully entered modernity. Europe witnessed another scandalous concert before Stravinsky's, in Vienna, where Arnold Schoenberg abandoned traditional musical form. In the United States Louis Armstrong played the saxophone for the first time and hinted at sonic revolutions to come.

The visual arts were at a pivotal point. The International Exhibition of Modern Art (the Armory Show) in New York introduced America to modern art. Kazimir Malevich launched his Suprematist Manifesto at the '0,10' exhibition and unveiled the painting *Black Square on a White Background*. This kind of total abstraction was startling: 'It is an end point for art – and yet, at the same time, the beginning of something completely new.'[5] Later in 1913 Marcel Duchamp's *Bicycle Wheel* was an equally audacious move, the first 'readymade' art. In the space of twelve months the full reaches of abstraction had been breached.

Adolf Loos published an architectural manifesto, based on an earlier talk, 'Ornament and Crime', setting the template for a clean, simple architecture that has never gone away. Walter Gropius, founder of the Bauhaus School, and with Loos another central figure in twentieth-century architecture, published his essay 'The Development of Modern Industrial Architecture'. Stark lines, since dominant, had their genesis here.

Speaking of publication, there came a welter of important novels. James Joyce sent the first chapters of *A Portrait of the Artist as a Young Man* to Ezra Pound; Virginia Woolf delivered her first novel to her publisher. D.H. Lawrence published *Sons and Lovers* and Marcel Proust the first volume of *A la recherche du temps perdu*. Book by book, the possibilities of literature were being transformed. And not just literature, but thought. Ludwig Wittgenstein published his first paper in 1913 and was busy working on what would become one of the century's most important books of philosophy – the

Tractatus Logico-Philosophicus. Carl Jung had just published *The Psychology of the Unconscious* (outlining the collective unconscious and causing the first major schism in the new discipline of psychoanalysis) and Freud published *Totem and Taboo*.

Science too was caught up in revolution. Albert Einstein walked with Marie Curie in the Swiss hills, broaching ideas that had first occurred to him in 1907 and would eventually become his General Theory of Relativity, his extraordinary idea of curved spacetime. In one of the most significant insights in the history of science, Niels Bohr provided the link between physics and chemistry, confirming the earlier instincts of Max Planck. There was a sense of a new world being built, a world in which Charlie Chaplin appeared in his first films and Henry Ford introduced the production line in Detroit.

Hanging over the tumult is the First World War. But it's striking how much was new in every corner of endeavour, from narrative films to the birth of Expressionism. The structures, foundations and categories of thought and art were being exploded, reimagined from the ground up in ways inconceivable even a generation before. *The Rite of Spring* is emblematic of a year when it was thrillingly possible for big ideas to change the world. This was of course the period of the 2IR – yet it wasn't just a moment of industrial revolution. It was a revolution of the human mind and imagination.

And it wasn't just a flash in the pan. The legacy of precocious invention lasted.

Living in the twentieth century, you always heard something new. Whether John Coltrane or John Cage, Joni Mitchell or Johnny Rotten, rock and roll or hip-hop, Schoenberg, the Velvet Underground or Björk, death metal or minimalism, gabba techno or the blues, the whole period was an electrifying auditory rollercoaster. Time and again, music would disrupt itself in the most public and comprehensive fashion. As soon as anyone was comfortable with a genre or style, new instruments, techniques and attitudes would explode from the underground into public consciousness. From Stravinsky on, music began to define and drive rebellion. Each generation dutifully hurled pathbreaking ideas about sound and culture into the face of an incredulous audience.

Does this still hold? Much less so, if at all. I love music, and in particular new music. I listen to it every day. But that doesn't mean that the great aural pattern of the twentieth century works on unimpeded. Arguably the last great musical revolutions were the growth of hip-hop and its subfields, and then the great variety of electronic genres. Within each there is still an extraordinary amount of elaboration: drill or trap or any number of permutations on the main theme. But arguably they are variations of established modes in a way that, say, rock or jazz were not. The paradigm shift has already occurred.

Almost everything we listen to now could have been produced twenty years ago. For most of the twentieth century that wasn't true. The music of the 1980s would have been inconceivable in the 1960s. Ditto that of the sixties in the forties. Jimi Hendrix was a world away from Vera Lynn, just as Madonna was a world away from Hendrix. The signature stars of the present and twenty years ago are, meanwhile, on a continuum, if not identical then hardly reinventing the nature of sound and style. All of this takes place in the context of the overall increase in production. Music and the creative arts have more opportunity and output than ever before. Forty thousand new songs are added to Spotify every single day. Nearly four million creators engage with the platform.[6] Never before have we had so much new music, or so many talented people producing it. And yet arguably we do not live in such a musically exciting time as preceding generations. We're in a rolling cycle of refresh and marginal innovation.

The writer Kurt Anderson makes the point that what applies to music, applies to almost every aspect of our culture. Whereas almost any point in the twentieth century was instantly distinguishable from moments twenty years either side, that is no longer true in fashions, films, architecture, novels, design or art. In his words, 'during these [last] 20 years, the appearance of the world (computers, TVs, telephones, and music players aside) has changed hardly at all, less than it did during any 20-year period for at least a century. The past is a foreign country, but the recent past – the 00s, the 90s, even a lot of the 80s – looks almost identical to the present.'[7] This is culture as *Groundhog Day*: stuck on repeat, the same TV shows recycled on schedules, the same architectural motifs and materials, the same athleisure fashion items season after season. The great movement of underground to mainstream seems to be short-circuiting in an age of streaming, instant celebrity and YouTube. To paraphrase Ross Douthat, this is not culture as a vital workshop, being constantly forged anew; it is culture as a museum through which we absentmindedly browse, over and over.[8]

Much of what formerly constituted counterculture is now vapid, toothless, deracinated. Cultural rebellion has been neutered, monetised and tamed for a global audience in a theme park version of public extremity. Music isn't the heartland of a broiling subculture so much as an accessible, easily digestible menu to be sampled from at will. Films are franchises. Original ideas face box-office death. Between 1980 and 2000, 305 of the 400 top twenty films every year were originals – that is, not a sequel, prequel, reboot, franchise rollout or spinoff. Between 2000 and 2020 only 189 of the same total were original; as time went on the number of originals in the top ten shrank to virtually nothing.[9] (Such 'originals' still include all the films based on books, TV and video games, as well as the first in a franchise.) I like Marvel and *Star Wars* as much as the next fan, but the era of art or culture as the vehicle for the shockingly new is over. Rather than the underground we have the blockbuster; not David Bowie but Ed Sheeran. The avant-garde that drove so much, challenging every preconception of art and creativity, has fallen apart.

On every front we are more capable, have more people and tools and brain power, more creative energy and output, than was available in 1913. And yet there is a sense that, with a reduced tolerance for aesthetic or intellectual risk, the twenty-first century is in no way better at producing big ideas than the twentieth. In intangible areas like music, we are caught in the same mixed pattern of the breakthrough problem and diminishing returns at the frontier: manic general idea generation coupled with the slow death of the big idea. The pattern has metastasised into a broader cultural phenomenon.

The invention of perspective; the Shakespearean soliloquy; utilitarian ethics; *Half-Life* 2: big ideas are waypoints of human experience that reach far beyond what is easily measurable. Most of our big ideas defy quantitative appraisal: the wheel, monotheistic religion, Renaissance humanism, democratic government, quantum theory . . . But you can still gauge qualitatively the general pattern of ideas.

To be sure, such ideas still exist: computer games, for instance, are a whole new field of cultural and entertainment experience, one whose rapid development and propensity for experimentalism make it a particularly fertile source of new ideas. In general though, we are not better at big new artistic ideas, even if we produce many more works of art. We've scaled up output, just as we've scaled up car production. Yet, just as we don't have flying cars, we have a safe and predictable culture.

It's actively hard to relive a time when configurations of brushstrokes, unusual words or music notes could viscerally shock us. Most of the major storytelling genres in books or films, for instance, haven't changed in decades. Changes are incremental rather than the sensory assault of modernism, with its ambition to overturn every aspect of aesthetic experience.

Perhaps there is a connection with an observed decrease in creativity. Since the 1990s measures of creativity among US students have been falling, sparking talk of a 'creativity crisis'.[10] These measures include skills like original thinking (the ability to generate new ideas, and in great numbers) and also any decline in open-mindedness; that is, less of a creative attitude and a welcome view of new and different ideas. This is likely part of the broader dynamic that inhibits many forms of radical originality. It doesn't bode well for a new renaissance.

In the realm of pure thought, the home turf of big ideas, we don't fare much better. It's striking that much current moral, theological and philosophical debate stems from the 'Axial Age' of the last millennium BCE. We still follow the religions of that era: Buddhism, Jainism, Zoroastrianism and Christianity. The axial sages of both Europe and China cast long shadows: Confucius, Mencius and Laozi; Socrates, Plato and Aristotle. This era of prophets and philosophers also laid the foundations of science, logic and biology. In matters of art or thought, big ideas are not rendered redundant as they are in technology; talking of 'progress' here can be redundant or anachronistic. But limning those long shadows puts the contributions of our own time in context: two thousand years of constant endeavour doesn't automatically get you big ideas beyond a Confucius or Socrates. And by implication, several decades more may not be sufficient either.

Consider how many fundamental aspects of our intellectual lives date to the nineteenth century and earlier: nationalism and the nation state; the capitalist corporation and industrial production; mass culture and professional science, to name just a few. From around the middle of the nineteenth century we had an explosion of big thinking but also a great undoing, what Felipe Fernández-Armesto calls 'a graveyard of longstanding certainties'.[11] Even as fundamentals dropped into place – for example, the laws of thermodynamics and Darwinian natural selection both date from then – it wasn't long before a grand unravelling kicked in.

Progressively, over the twentieth century, big idea after big idea dismantled prior certainties. Almost every aspect of thought underwent a paradigm shift; creative conceptual destruction, disruptive epistemic innovation, became the rule. Einstein and Henri Bergson rebuilt time, while everyone from Wittgenstein and Ferdinand de Saussure to Jacques Derrida turned language from a medium of truth into a hall of broken mirrors. But this was only the beginning. National culture, the self, art, logic, the foundations of mathematics, religion and nature, the knowability of reality, hierarchies of class, gender and race, the history of the universe, the dawn of the human species, the assumed fixity of continents, structures of meaning: all were torn apart in their traditional guises by a bravura, decades-long firework display of radical thought. The intellectual revolutions of the twentieth century remade everything.[12] Big new ideas ripped apart the foundations of understanding. But as the historian of ideas Peter Watson points out, it was also a time of extraordinary invention: around 1900 alone such discoveries as the gene, the electron, the quantum and the unconscious were made.[13] Across the board came a bold parade of 'isms', from Abstract Expressionism to behaviourism to Cubism, and on and on.

Such an era of big ideas is, if not finished, de-energised. We aren't producing new thinking of this level at anywhere near the same scale. One symptom is the 'death of the intellectual'; the dearth of people whose basic function is to conjure and communicate big new ideas.[14] The intellectual as a self-conscious and serious category is faltering, even in their heartland: the Left Bank of Paris. From eighteenth-century *philosophes* like Voltaire to heroic writer-thinkers like Sartre, from Émile Zola to Michel Foucault, there was a French tradition of unashamed breakthrough thinking. But even in the Latin Quarter the intellectual is now a nostalgic figure, not so much a contemporary force as a caricature with a black turtleneck and a pipe. Intellectuals, like creative artists, have seen their rebellion defanged. For many, the notion of having a big idea and communicating it directly to a wide general public is inconceivable: they are narrowly specialised, communicating to one another in obtuse academic vernacular. For others, the promise of quick celebrity and a TED talk does quite the reverse. Meanwhile big ideas have become gauche, unfashionable, seen as risky, wrongheaded and unwieldy.

There are almost certainly more academics alive in the last twenty or thirty years than at any other time in history, working across the full span of intellectual endeavour on political theory, philosophy, anthropology, sociology. But that great mass of scholars burrow down into niches and publish in little-read journals. The shape of academia is more settled as a result: in the twentieth century we gained such major new disciplines as anthropology and psychology, computer science and biochemistry, management studies and media studies. In the twenty-first century came a proliferation of niche masters' degrees, but few major new branches of knowledge.

Few would call this a golden age for original thought. Bemoaning the state of the humanities and social sciences, the anthropologist David Graeber saw an attenuation of ambitious thinking. Everyone still endlessly discusses the thinkers of the 1960s and 1970s without producing comparable work: 'No major new works of social theory have emerged in the United States in the last thirty years.'[15] The philosopher Agnes Callard agrees, writing, 'When I am asked for sources of "big ideas" in philosophy – the kind that would get the extra-philosophical world to stand up and take notice – I struggle to list anyone born after 1950.'[16] Similarly a study of 500 Western polymathic intellectuals finds plenty born in the 1940s, from Julia Kristeva to Vaclav Smil, Jacqueline Rose to Bruno Latour, but encounters a precipitous fall from the 1950s on: 'The drop around 1950 may be an alarm signal', the author writes.[17]

In a core area of big ideas – political thinking – intellectuals are in full-scale retreat. As the twentieth century wore on, and especially with the collapse of the Soviet system, ideology, and the idea that you could invent and pursue a new ideology, faded. Whether you were in China, Russia or the USA, politics became a matter of getting things done within the system. There is no twenty-first century version of Marxism or capitalism or liberalism or even something like anarchism. There are just shades of grey, old, regurgitated political ideas, a pre-existing menu.

For example, no one really has a concrete suggestion for what, after the collapse of communism, might advantageously replace capitalism or liberal democracy as the dominant principles of world organisation, or even whether this is possible. Concepts like democracy and capitalism are commonly assumed to be in crisis. Yet without clear directions for the future or any replacements, they lurch on.

In the late 1980s and early 1990s, Francis Fukuyama

noticed the phenomenon, famously (and more optimistically) calling it the 'End of History'. His hypothesis has been mocked as the epitome of liberal overreach, but it's also widely misunderstood. He neither believed nor argued that events, including those of the highest significance, would stop, just that the principles and institutions of government and politics were unlikely to develop much further. Fukuyama's reading of political philosophy suggested that a major evolution of economically efficient and psychologically satisfying capitalist liberal democracy was unlikely.[18] Had Fukuyama more prosaically called his article and book 'The End of Ideological Evolution and Political Big Ideas' it would, one suspects, have attracted less controversy. Moreover, Fukuyama never blithely assumed that liberal democracy was assured; not only might it fall prey to external threats, he pointed out, but liberal democracies could wither internally from contradictions, a lack of new ideas, misplaced focus, anomie, disillusion and distraction. Much of which we now see.

Are we to accept that we really have reached the end of big ideas, that in this regard we've arrived at the perfect settlement? Currently the only meaningful challenge is a Chinese-style autocracy, a managed capitalism that blends twentieth-century authoritarianism and neoliberalism in a strange and seemingly powerful brew. There is a kind of global political and ideological fudge between liberal democracy and what Fukuyama calls the bureaucratic authoritarian states; they mark, if not the 'final form ... free from contradictions', the present de facto limit on political ideas.

There is no radically new or ambitious imagining of the world that doesn't replay existing concepts. For the most part programmes, policies and ideologies aim to adapt or perfect the present system. We may live in turbulent times, but this takes the form of reversions: authoritarianism, strongman tribalism, sectarianism and nationalist populism are hardly new or original ideas, even if they have been reheated with digital and globalised twists. Reversion to identity, whether racial, religious or national, is everywhere from Myanmar and India to the United States, Brazil and Poland, just as countries from Turkey to Russia cleave to a reheated authoritarianism. Just as there is nothing new about capitalism, religious fundamentalism and neo-fascism are hardly original.

But no one here is trying to reinvent political theory. No one is claiming their authority rests on something other than the interests of the people, even if there is debate about what that involves in practice. No model suggests an unseen direction for further social or political evolution, a continuation of the great dialectic of history.

Basic political categories – left and right wing – date back hundreds of years. In general we remain trapped in a rehearsal of the moulding dichotomies of markets or states, Hayek or Keynes, individuals or collectives, socialism or capitalism. Political tropes, like their cultural counterparts, are endlessly recycled, whether it is the feckless poor or greedy financiers. Most radical policy proposals are decades old: nationalisation; privatisation; fiscal tweaking with tax rises and tax cuts. This paralysis means 'new anxieties' are faced with 'old ideologies'.[19] Some new ideas get through: nudge unit governance for example, or perhaps nascent concepts like Universal Basic Income or Modern Monetary Theory.[20] But in general, despite progress, we lack ideas to solve questions as diverse and significant as loneliness, homelessness, social care, care of the elderly and child care.[21] Even the most energised debates – about race, gender or the environment, for example – have their roots in the 1960s and earlier. The Covid-19 pandemic was, in its early phases, thought to be changing every facet of life; its impact has indeed been enormous, but this has mainly involved accelerating pre-existing changes like uptake of digital technology. Perhaps the most surprising thing is how little has changed structurally. Events unfold, yes; but change at the frontiers of political economy is glacial.

Yet there has been a revolution in policy ideas (let alone ideologies) before. At the beginning of the nineteenth century, governments were about 'laws and wars'. One hundred and fifty years later, they engaged with almost every aspect of their citizens' lives, from museums and culture to birth and old age.[22] Especially between the late nineteenth century and the 1960s, there was a wholesale revolution in thinking about the state. From surveying poverty to state pensions to the development of a fully fledged system of social security and public health, the aims and nature of policy were rewritten. Governments built new towns and hospitals; provided humane cover for the unemployed or elderly; expanded education at all levels and funded research. Is that kind of bold policy agenda definitively over?

Some might, with justification, say good riddance to ideology and intellectuals alike. For decades Maoism held a bizarre stranglehold on Left Bank philosophers, despite killing tens of millions in a needless famine and embarking on one of recent history's most deranged and destructive episodes in the Cultural Revolution. Mussolini, Stalin, Kim Il-Sung, P.W. Botha and Pol Pot arguably all had big political ideas, and we certainly don't want them back. Nor should we deny our undoubted political and economic achievements: colossal growth, marked improvements in most areas of life, fairer, more open polities across much of the planet, a previously unmatched availability and quality of cultural and artistic experience. Good points, but surely the challenge is to not only bank gains but envision a better world still.

So it's not all doom and gloom. But as Yuval Noah Harari

among others has pointed out, there has been a breakdown of grand narratives (religious, political, cultural) as to our purpose as individuals or societies or a species, without a corresponding reinvention. In areas from policy to architecture, music to philosophy, big ideas have withered. Talk of the human frontier here is always debatable. And yet, in everything from the films we watch to the political organisation of society, I'd argue that the frontier is a more settled, less dynamic place. We are good at producing things, we are not nearly as good at thinking anew. Except, surely, in one area, perhaps the most fundamental of all.

The Science of Science

The body of knowledge and methodology that constitutes science is a good candidate for humanity's greatest achievement. Humanity works with scales from the Planck length to the observable cosmos. Our minds can cycle back billions of years to the churning inflationary birth of the universe. We can unlock the chemical foundations and internal structures of inordinately complex organisms. We have a powerful, proven description of nature. From gravitational wave astronomy to the detailed observation of exoplanets, from the biochemistry of new diseases to molecular engineering, it's not hard to find examples of stunning successes and discoveries. Look at any given month of recent years, and you find significant advances across the board, extraordinary feats of engineering, probes to the farthest reaches of existence or life. Historical claims that science has hit a wall have always disintegrated in the face of a new round of discovery.

And yet there are still questions.[23] The focus of research has often shifted to the incremental and safe, even as much contemporary work is 'contestable, unreliable, unusable, or flat-out wrong'.[24] In recent decades there has arisen a palpable feeling that the era of great leaps and revolutions has passed; that part of the reason we face technological stagnation is that its scientific interlocutor is also hitting headwinds. Science is not only caught up in the dynamic of consuming ever more resources to maintain a steady and at times frustrating rate of progress, but could arguably be its source.

The case goes something like this: the outputs of science, usually research papers, have been growing at an exponential rate for decades. Depending on the discipline, their numbers double every ten to twenty years.[25] Growth in journals output is around 8–9 per cent a year, implying an ever faster rate of doubling – every nine years.[26] In biomedical research the PubMed database sees over a million new papers a year;[27] 12–13,000 papers are uploaded to arXiv, the main repository of physics research, every month, up from around 2000 twenty years ago.[28] According to one estimate, Google Scholar incorporates nearly 400 million documents.[29] Indeed, scientometric research indicates that such growth is part of a centuries-long pattern of exponentially increasing knowledge production.[30] More science PhDs are awarded and more funding granted than at any previous point, and by a decisive margin. Graphs of all three – papers published, PhDs awarded, funding dollars – rise vertiginously from the postwar period to the present. In one large research university today, there may be as many researchers working in a field as in the entirety of Europe or America a hundred years ago.

We should expect extraordinary results given this output, the scale of which dwarfs anything in history. But while the growth in numbers of papers may appear to be exponential, analysis of the literature indicates that new ideas only grow in linear fashion.[31] For every new idea, then, the number of papers produced is now much, much higher. This is the same pattern we saw with new technologies, and no doubt would hold for a similar analysis in the cultural field, with films or novels.

Scientist Michael Nielsen and entrepreneur Patrick Collison have surveyed leading scientists to see what difference this has made.[32] They asked them to compare Nobel Prize-winning discoveries: which was more significant, say, the discovery of the neutron or cosmic background microwave radiation? Assess a large enough sample of these comparisons (they looked at 4483) over enough influential scientists and you should have a reasonable picture of how science understands its own progress; that is, whether a discovery in a given decade is ranked as more significant than those in other decades.

In physics the period between the 1910s and the 1930s was a clear high point, a golden age which established the foundations of the modern discipline. But, with the exception of a spike in the 1960s (which saw the discovery of cosmic background microwave radiation and the Big Bang, and the establishment of the standard model of particle physics), that significance has never been recaptured. In physiology and chemistry there are perhaps slight upward curves, but those are small and there is still a lot of variance over the decades. Across disciplines there was a decreased tendency to award prizes for recent work compared with previous decades.

Nielsen and Collison cite two modern examples as indicative: the discovery of gravitational waves and the Higgs boson. While both are fantastic, compare them to their originating theories. In 1915 Einstein's theory of general relativity 'radically changed our understanding of space, time, mass, energy, and gravity'.[33] Instead gravitational waves corroborate it; we are still confirming the predictions of a century ago. Likewise, the Higgs boson fits into a pattern of particle discovery that, in the twentieth century, populated a hitherto undreamt-of array of particles, from neutrons to antimatter. In 2013 just two people, Peter Higgs and François Englert, won the Nobel Prize for the discovery of the Higgs boson. Yet 3000 people were named as authors on the key papers that went into finding it. CERN spent some $6 billion building the Large Hadron Collider, a major infrastructure with thousands more workers, builders, engineers and support staff. One or two people are hardly responsible for the discovery; all that activity is built atop fairly old predictions from individual minds.

Expenditure, numbers of PhDs and publications have all grown between ten- and a hundredfold; yet scientific progress and significant discoveries, in the eyes of its most important contemporary exponents, haven't. Scientific big ideas are also in a long-term pattern of diminishing returns.

No one doubts that science will continue growing, particularly in applied fields, even if the impact might wane. What is at stake is the ability of science to revolutionise itself with profound shifts of view. Will we have another discovery with the significance of heliocentrism, Darwinian natural selection, Mendel's genetics, quantum mechanics, Bohr and Pauling's unification of physics and chemistry, the structure of DNA, the Big Bang? Some, like the science writer John Horgan, believe this is unlikely, although that claim remains controversial.[34] The making of such revolutions will however get more difficult and require more input. They are by no means guaranteed.

This is borne out by the conduct of science today. Before the Second World War the phrase 'Big Science' would have been meaningless. Today it is routine. A study in *Science* found that in 1955 around 50 per cent of engineering papers were team authored.[35] By the 2010s, this was 90 per cent. Today a team-authored paper is six times more likely to receive over one thousand citations. Team size growth is constant, averaging 17 per cent per decade. Similar outcomes are to be found in most scientific disciplines. In the last decade the number of papers in the Nature Index with more than one thousand named authors increased from none to well over a hundred, while Web of Science records more than a thousand papers with over a thousand authors published in just the years 2014–2018.[36]

Again and again, the pattern of the Higgs boson is repeated: what took an individual or small team to begin requires thousands to finish. The structure of DNA was discovered by essentially three people: Watson, Crick and Franklin. When it came to one of the undoubted breakthroughs of recent times, the sequencing of the human genome, it took thirteen years, multiple global teams working with and against each other and, conservatively, thousands if not tens of thousands of highly trained scientists. The pattern holds in nuclear energy: from Rutherford's discovery of atomic structure almost alone on a basic Cambridge workbench to the awesome scale of the Manhattan Project. Or think of Einstein predicting gravitational waves and then the decades-long multi-billion-dollar endeavour of confirming their reality.

The direction of travel for large portions of research is towards ever more sizeable teams and experiments. This suggests that delivering big ideas in science is harder than ever. (And now there is a growing body of evidence to suggest that small teams are more innovative and capable of disrupting science.[37]) Furthermore, that direction of travel is down a more specialised, narrower road. It should be no surprise then, that despite all our advances we have Eroom's Law, or cannot easily build new forms of transport, or change our society.

We have seen that, compared with previous eras, the scientific progress we do have translates less readily into economic growth and productivity improvements.[38] The building blocks of knowledge keep growing, but they are heavier, more unwieldy, hidden down rockier paths than in the past. The implication is that, in science as elsewhere, the realisation of big ideas will need ever greater efforts over time.

As we will discuss, this hunger for resource opens science to a plethora of financial, social and organisational challenges. In turn this leads some in the scientific community to worry. They feel it renders Big Science conservative and predictable, subject to gaming and perverse incentives. Is there still room, to paraphrase Freeman Dyson, for the scientist as rebel? Competition for roles and resources is intense and draining. The field becomes ever more gerontocratic: for example, National Institute of Health grants for young people have collapsed. In 1980 21 per cent of NIH grants were made to teams with a principal investigator under the age of thirty-five.[39] By 2014 it was around 2 per cent. Meanwhile the share of principal investigators over the age of sixty-six grew from about 1 per cent to nearly 10 per cent. We can't draw too many conclusions from age profile alone, but it doesn't look like a system priming itself for disruption.

Science faces dissent from within, symptoms of trouble at the frontier. There is an uneasiness in some quarters that, especially at the limits, theories are becoming unfalsifiable, 'post-empirical'; that without a plausible experimental test they no longer constitute science. Physics is actually full of big ideas, arguably the biggest ideas around. String theory and its evolution to what Edward Witten called M-theory are great examples: massive constructs of thought and mathematics, elegant, intricate, vast, built since the late 1970s and early 1980s – but not experimentally confirmed. Theories of brane worlds and the multiverse rub shoulders with loop quantum gravity, eternal inflation and many other exotic and enthralling ideas. Areas like particle physics and cosmology hence face a paradox: home to many of the biggest, most exciting ideas around, they often lack the kind of decisive verification science used to specialise in. The existence for decades of multiple theories is a rarity in the history of (modern) physics. Again, the conception of big ideas is not the problem; rather advances get stuck at the execution or purchase stage, creating what the philosopher of science Imre Lakatos called a 'degenerating' research programme.[40]

Separately, in discussions of contemporary science, one phrase recurs with wearying frequency: reproducibility crisis.[41] If the results of experiments (and social scientific papers) cannot be replicated, and this is happening at scale, then crisis is not an ill-chosen word. A *Nature* poll of 1500 scientists found that 70 per cent had failed to reproduce results.[42] In some fields up to 50 per cent of studies fail to replicate.[43] The figures compound corrosive evidence of methodological error like small sample sizes, inaccurate statistical analyses, bias and even fraud in much scientific work, a system in which the gaming of results for maximum impact trumps rigour.[44] What's more, as in cancer research, the accumulation of science generally creates its own problem: there is simply too much relevant material for scientists to consume. Important, often older work is thus regularly missed.[45]

Some, like the physicist Lee Smolin, maintain that progress on the most fundamental challenges has been particularly disappointing. Questions like the relationship between quantum theory and gravity, or the nature of dark matter and dark energy, remain unanswered. He argues that from the end of the eighteenth century physics leapt forward in twenty-year intervals. This progress has stopped: 'since the end of the 1970s there has not been a genuine breakthrough in our understanding of elementary-particle physics.'[46] Sabine Hossenfelder is blunter still: 'Nothing is moving in the foundations of physics,' she argues. 'The self-reflection in the community is zero, zilch, nada, nichts, null.'[47] More diplomatically, as *New Scientist* puts it, 'none of the really ambitious ideas of the last 30 years or so have come good'.[48] In hyped areas like neuroscience results have been mixed, and have arguably not provided the step change trailed by the discipline's excitable boosters.[49]

The extraordinary progress that has been made leaves a raft of lacunae: it hasn't confirmed or excluded the possibility of alien life, a question of cosmic significance, let alone fully accounted for the origins of life on Earth. It hasn't solved the puzzle of consciousness or reversed the ageing process. Even asking these questions is verboten in many circles.[50] It hasn't uncovered the ultimate building blocks of spacetime, hit upon a watertight 'theory of everything' (or agreed what that means) or fixed limits to our control of nature. There is no necessary reason why it should of course, except that we grew habituated to an over-achieving science. But it is worth remembering that such mysteries remain.

There is also a danger of hubris. Science is supposedly built around an organised scepticism; but there is a chance that it has calcified into arrogant scientism, buying too heartily into its own dogmas; cutting itself off from strange or seemingly unproductive avenues for reasons of bureaucracy or narrow-mindedness.

As I write, I've been leafing through a recent edition of *Scientific American*. It contains multitudes: the largest ever experiment to detect neutrinos, experiments to find new particles and identify dark matter, pioneering research on supermassive black holes in the early universe, explorations inside neutron stars and studies of quantum entanglement. It doesn't feel like a struggling area, and that isn't the word I'd use. Instead science is both an inspiring success and also subject to increasingly potent diminishing returns, especially in its most radical and fundamental guises. The 'river of discovery' flows on, but its character has changed: the river itself has never been larger or absorbed more, but the discoveries within it grow, in comparison, smaller.[51]

Like other endeavours at the frontier, science is caught in a series of contradictions that are shaping its future, just as science and its own contradictions will shape the future at large. The real significance is not that science flourishes or stalls; it is, unlike in other eras, the complex, contradictory nature of a time where either interpretation is possible.

Rational Pessimism

Today, as for nearly a thousand years, my hometown of Oxford is dominated by its eponymous university, a sprawling institution that effectively forms the city's central district. Much about the university gives a sense of permanence and timelessness: hushed sandstone quadrangles draped with ivy; begowned dons conducting esoteric rituals, graces and arguments in wood-panelled rooms; confident undergraduates living their languid Brideshead fantasies. Much as anyone connected with the university tries to deny it, the clichés are true.

But over the last two hundred years, the university has undergone an astonishing transformation. In the eighteenth century Oxford was an intellectual backwater. Despite setting the pace of medieval philosophy and theology and playing an important role in the Scientific Revolution of the seventeenth century, it had settled into a cosy torpor: part Anglican seminary, part finishing school for England's landed elite, it was dogmatic and obscurantist.

Change was coming. A set of enterprising Victorians put the university on a modern footing. Spurred on by the German idea of universities as research powerhouses, Oxford clawed its way back to the frontier. By the 1930s it was shedding dilettante amateurism and hosting a series of world-changing breakthroughs. It was helped by a generation of émigrés including major physicists Einstein and Erwin Schrödinger and classicist Eduard Fraenkel. As we saw, Howard Florey and Ernst Chain carried out their pivotal work on antibiotics, arguably the most important medical advance of the entire century. Dorothy Hodgkin began pioneering work on X-ray crystallography that would eventually win her the Nobel Prize in Chemistry, a tool later pivotal in a further twenty-nine Nobel-winning projects to date (two other Oxford chemists from the same era also won the award).[52] Oxford quickly became a leading centre for the sciences, a role traditionally held by its great rival Cambridge. Its researchers won prizes and plaudits.

Social science was not forgotten and Oxford's traditional strengths in the arts and humanities were burnished. Literary academics including J.R.R. Tolkien and C.S. Lewis formed a group called the Inklings – and in doing so, did as much as anyone to create the modern genre of fantasy, a pillar of today's entertainment industry and imaginative universe. With its distinctive brand of linguistic philosophy and names like Ryle, Ayer, Austin and Strawson, its philosophy department became arguably the most influential in the world. After centuries of scholasticism, Oxford was once again at the global epistemic frontier.

That is a role Oxford maintains. But it has since won fewer Nobel prizes: none, in fact, between 1963, when Dutch animal behaviourist Nikolaas Tinbergen won and 2019, when Peter Ratcliffe won the Physiology prize for his cellular studies of hypoxia (the great mathematician and physicist Roger Penrose won a prize the following year). And yet all its key metrics are up. Its annual income tops £1.6 billion, much more if you include the turnover of its publishing business, a sixty-fold increase in real terms on the figure of the 1930s. With 1817 academics, there were three times as many. The number of research-oriented postgraduate students jumped twenty-fold from 536 in 1938 to nearly 11,000 in 2016.[53] Postdoctoral researchers didn't really feature in the 1930s, and even into the 1970s they numbered in the low hundreds. Forty years later Oxford had around 5000 postdocs, all devoted primarily to research.

Likewise, research grant income was barely a category in the 1930s. From 1922 the British government gave the university £100,000 a year for science research. By the late 1980s its research income had reached £40 million per year, and by 2016 it was £723 million and rising.[54] Its Medical Sciences Division, ranked as the world's best, would comfortably be one of Britain's largest research organisations on its own. All measures of research output have risen steeply since the 1930s and particularly since the 1970s and 80s. In 1939 the university employed only 200 non-academic staff. Now there are 6665, and a similar number are employed by the colleges.[55] There is a much-expanded real estate profile, and a greatly expanded range of tools: powerful computing and experimental equipment are ubiquitous, as they are at every leading research centre. The university is capable of raising billions on the bond markets to fund further expansion.

This shows in miniature the strange state of big ideas in the twenty-first century. Oxford has never had it so good. And with its Covid-19 vaccine, it has scored a medical research triumph. Beneath the fusty veneer is an organisation experiencing meteoric growth across all its key measures. It has never produced so much academic work, of such high quality. Yet it has taken eighty years for Oxford to match the discovery of penicillin. And why doesn't Oxford win more Nobel prizes when, on the face of it, it is so much better equipped?

With respect to the biggest ideas of all – whether in medical research, philosophy or English literature – it has made little more impact than in the 1930s. When it comes to one university at least, there seems to be diminishing returns. You could say that other places have improved much more rapidly – but Oxford is ranked as the best or close to the best university in the world. It is a symptom, not an outlier.

Diagnosing stagnation is difficult. Not only is there no agreed-upon measure, but we would also expect there to be decade-by-decade fluctuations in revolutionary ideas anyway. Perhaps all this is just 'normal' turbulence, rather than indicating a deeper underlying problem. There is no neutral position from which to decide. Perhaps society has become accustomed to the slower or less impactful delivery of less impactful ideas – so prevalent it takes on a 'creeping normalcy', hidden in plain sight as year after year we unwittingly readjust our expectations.

The school of rational optimism builds a powerful case that the world is, in fact, drastically improving. They argue – rightly – that on average, humanity lives longer, healthier, wealthier, safer lives than ever before.[56] The scientist David Deutsch argues we are at the 'beginning of infinity', an optimistic and radical explosion of ideas and progress, founded on opening new forms of knowledge and understanding. How does that picture square with the argument for a slowdown in the production of big ideas?

I'd argue the place to look is not just at where we are, but where, given everything at our disposal, we *could* be. That should help answer the above question – and prompt a dose of rational pessimism.

Today there have never been more people capable of adding to the sum total of human achievement. We are more numerous and better educated than ever. School enrolments have risen steeply. In 1900 just 6.4 per cent of Americans completed high school, but now well over 90 per cent do. In the US, educational attainment increased by 0.8 years per decade over the period 1890–1970.[57] Moreover, on certain standardised measures evidence indicates that the world is getting cleverer. Since the 1920s the world's population has gained just over twenty-five IQ points, more in some countries.[58] Such increases are concentrated in areas like analytic and abstract reasoning. People today also have more leisure time to pursue their interests: thanks to shorter working hours, more holiday and earlier retirement, the amount of life we spend working has gone down by a quarter since 1960.[59]

Oxford is not alone. The entire UK university sector has seen huge growth: from a small number in the postwar years, today there are 168, supporting half a million jobs and receiving over £35 billion of annual revenue.[60] They have never been so generously endowed with books, labs and well-trained researchers. Speaking of endowments, at least twenty-three elite US universities each have more than $5 billion in the bank and over a hundred sit on more than $1 billion.[61] The US higher education sector has revenues equivalent to economies like Saudi Arabia and Argentina.[62] Around 7.8 million researchers work in professional science, by far a historic record.[63] Other estimates suggest that 90 per cent of the scientists who have ever lived do so right now.[64] Over the twentieth century, the number of knowledge-producing workers in the US increased by a factor of nineteen.[65] Nor does this general boom only apply to STEM (science, technology, engineering and mathematics) subjects: creative arts and humanities also saw large increases in numbers of students and practitioners, and in their exploratory capacity.

Creating new ideas requires resources. Typically the

availability of resources is described by national R&D figures, although they almost certainly understate R&D in areas like the arts and creative industries. Here again the picture is good. Estimates for global R&D spend vary, but two things are clear: first it is increasing, and second the totals are historically large. Global R&D spend is at least $2.2 trillion.[66] In 1973 the equivalent figure was just $100 billion; in 2000 $722 billion.[67] The share of R&D as a proportion of global GDP has ebbed and flowed around 2 per cent, but is now lodged above 2 per cent (around 2.3 per cent per year, according to the OECD) and as the economy grows, so does this expenditure.[68] The top 1000 corporate investors in R&D have a trend R&D growth rate of 4.8 per cent per year, putting in over $700 billion across sectors including IT software and hardware, healthcare, aerospace and defence, automotive, chemicals and energy. At this pace, spending doubles every fourteen years.

These are fecund times by many measures. Wikipedia alone is a marvel of the ages. The complete *Encyclopædia Britannica* runs to thirty-two volumes. Set out in the same way, Wikipedia extends to 2818 – and counting. More of everything is produced every year. Today 11,000 feature films are produced each year, 6 million new songs, 2.6 million new scientific journal articles.[69] Culturally speaking, our surfeit is almost embarrassing. Every year the Internet deals with over 2 trillion search enquiries.[70]

Computational power has increased – wait for it – a quadrillion-fold. Previously impossible calculations and tools are now trivial, from eye-popping graphics to hardcore big data analysis. Tasks that would have taken multiple people weeks are performed in fractions of a second. Simulations, designs, prototypes, models: all are now easy thanks to this growth. The connective tissue of the Internet forms the greatest publication and collaboration tool in history, up there with and even surpassing the codex or the press. Given that ideas often come from serendipitous connections between people and information, no wonder the Internet was billed as the most significant event in the production of ideas for centuries.

In the opening salvos of the third millennium CE, deep trends crescendoed. We became better educated and equipped, had more powerful tools, could gain access to the totality of human culture and achievement, invested more in all of the above and in endeavours pushing at the frontier. We inherited a world seemingly better attuned to both the random mashups of the ideational process and the painstaking work of reaching a Eureka moment.

All of this should raise expectations. Surely big ideas should arrive in an accelerating rush, accumulating ever faster, pushing at the frontier with increasing confidence. Like Herman Kahn and Anthony J. Wiener in the 1960s, surely we should expect great things and see massive breakthroughs not as miracles but as normal, a logical output from rising input. It's not unreasonable to expect a proportionality between the two. Likewise it is far from irrational to wonder why we have slower productivity growth, why recent research and technology yields less impact, why life expectancy, transport speed and a host of other measures from crops to computers to music and politics are not pushing back the frontier as they once did. The Great Stagnation debate is complicated, but its impact is real.

I wonder if the problem lies in the interpretation of ideas: a blurring of mass production with quality. The great category error of our time is to mistake extensive progress for intensive progress; the surfeit of new products, cultural outputs, things and services for deep, accelerating and underlying shifts. We've succumbed to a self-congratulatory parochialism: seeing the frontier buzzing with activity, but missing how little it achieves. This represents an historic decoupling between the ability to have new ideas and the ability to instigate big ideas; to change knowledge or culture or technology on the widest of scales.

Toolmaking hominids have roamed the Earth for something approaching three million years. Yet it is just in the last two centuries or so that we have experienced a sense of vertiginous acceleration. No wonder there is no clear-sighted perspective of our own place in intellectual history. Our entire worldview is framed by this step change. We take optimism, improvement and new big ideas for granted. We shouldn't.

This is not about doom-mongering or a hankering for a lost golden age. It's just facing up to the challenge of our time: not only barriers to progress, but the barriers to progress thrown up by progress itself. As we will see, the breakthrough problem is the fruit of great advances. What worked in the past cannot deliver indefinitely.

The frontier is at a crossroads.

Big ideas are conflicted. For most of human history, it would have been easy to answer the question whether there are likely to be more big ideas in the future. Usually it would have been a clear no. Stasis was endemic. Big ideas arrived at a rate of centuries or, optimistically, decades. Yet at other times, and particularly since the successive movements of the Renaissance, the Scientific Revolution and the Enlightenment, the answer would have been yes.

Our default is still set to yes.

In these last chapters I have suggested that it's not so simple. There is growing evidence that we are stumbling; that the paradigm won't always shift; that we need to spend more and more to maintain the production of new knowledges and technologies; that the institutions and people behind radical ideas may be less inclined to work on the frontiers. We are distracted by volume not quality, significance or bravery. We have private companies like SpaceX but precious little crewed space flight; we have the Standard Model of physics but haven't gone beyond it; we have libraries of humanities research that no one reads; we have a decades-long war on cancer, but still have cancer.

Our time comes with a litany of big ideas: blockchain, mobile social networks, supermaterials like graphene, deep learning neural networks, quantum biology, massive multiplayer online games, molecular machines, behavioural economics, algorithmic trading, gravitational wave and exoplanet astronomy, parametric architecture, e-sports, the ending of taboos around gender and sexuality, to name a few. But execution and purchase are more problematic than in the past. There is more evidence of struggle, sclerosis, decay and hesitancy than one might assume given everything at humanity's disposal.

The human frontier is in a curious place. Big ideas appear to be subject to diminishing returns, hitting headwinds that tell on their capacity to change our world. The early years of the twenty-first century are interesting because it is not wholly obvious whether we can confidently state that, in the future, our biggest ideas will be more abundant than today. For the first time in centuries, in almost any given dimension, we don't have an uncomplicated direction of travel. Rationally neither optimism nor pessimism makes sense any more. The situation is remarkably mixed for a society that *should* be firing on all cylinders. This is unusual. It is also of critical importance to our future.

Interlude

Enlightenment Then – A Big Idea in Practice

Paris, 10 December 1948
Thirty-five years on from Stravinsky's premiere, there was another big night in Paris. The newly minted United Nations did something that had seemed unthinkable just a few years before – its General Assembly adopted a Universal Declaration of Human Rights. After long months of wrangling, there wasn't a single dissenting vote. Humanity's moral frontier had been enlarged. One of the most significant ideas of all – inviolable, innate human rights – was given purchase.

Under the careful chairmanship of Eleanor Roosevelt, a motley but talented band had drafted an 'international bill of rights' fit for the new postwar world, reeling as details of Nazi atrocities became public. In thirty short articles the Declaration became a defining instrument of the twentieth century and beyond, a moral standard no one could entirely evade, adopted in at least ninety constitutions and legal systems.[1] It allowed the discourse of rights to move from philosophical discussion to the province of national and international law. It synthesised legal and ethical traditions from around the world: an Anglo-American tradition of individual liberty, but also Continental and Latin American traditions of a more active state, along with influences from Islam, Confucianism, Hinduism and Communism.

The proximate journey to the Declaration began in 1945 as the war drew to a close. At an international gathering in San Francisco instigated to set the foundations for the UN, it was agreed to create a Human Rights Commission, and to include mention of human rights in the UN Charter. But only in vague terms. Everything would depend on the Commission having the right balance of personalities and nationalities to find workable compromises. Above it all it would take an exquisite act of leadership to balance all the competing imperatives and traditions, and shepherd them through this nascent and often divided international body.

In January 1946, the UN General Assembly's opening session was held in London. There the Economic and Social Council asked Eleanor Roosevelt to convene a committee which eventually became the Human Rights Commission. The former First Lady was to be an inspired choice: just widowed, a campaigner, journalist and highly visible public figure, she was both committed to international co-operation and human rights and looking for a new role. Human rights were a progressive post-White House cause she could believe in, building on work she'd already done on racial and women's equality, workers' rights and international collaboration. Roosevelt was a member of America's first delegation to the UN, but as neither a career politician nor a diplomat she was unusual. She also had her doubts, not least that her independent-mindedness would conflict with strict State Department scripts. In the end, her desire to make a positive difference won out, while her global prestige and unselfish pragmatism was often able to carry conservative and hawkish US policymakers to uncomfortable places.

The Commission began meeting in New York State. Roosevelt was elected chair; the committee decided its main business would be drafting an international bill of rights (and thinking about how to implement it). Now with eighteen members, it started formal work in January 1947.

Ideas, histories and models, pre-existing rights instruments and constitutions, poured in and needed to be digested. But from the start everything was contentious. Communist and classical liberal (Western) perspectives clashed. 'Smaller' nations were adamant that their voice needed to be heard. Controversy and competing worldviews plagued the discussion, not least on the question of what, if any, social and economic rights should be included. Powerful nations were leery of any discussion of human rights on the grounds it might legitimate outside interference. At times it looked as if these fears might sink the whole project. It was never clear to what extent the instrument under discussion should be enforceable, if at all.

Even though a declaration wasn't legally binding, Roosevelt was convinced of its moral force. Whereas some wanted a meaningless proclamation and others a legally binding treaty or convention, she understood that a declaration would both carry weight, actually pass and keep the door open for future developments. Nonetheless, even if it wasn't enforceable, settling on the wording was never a done deal. Every word, line and article was debated, torn apart and put back together again, often over and over. Even with an array of talents and ideas at her disposal, it would take every ounce of determination to steer it to fruition. Although the Declaration was fortunate to have a group of philosophers-cum-diplomats instrumental to its delicate framing, including Charles Malik of Lebanon, Peng Chun Chang of China and René Cassin of France, Roosevelt's patience, perseverance, stature and mediating influence were critical.

Eventually, in late 1948 at the General Assembly in Paris, it was crunch time. Omens were disturbing. Although the drafters considered the document finished, a discussion of Article 1 alone lasted a week. Chair Charles Malik began to wield a stopwatch to keep contributions to a strict schedule before time ran out.

Eventually it did; there would be a vote. Presenting to the General Assembly in the Palais de Chaillot, Roosevelt pitched the document in a great tradition: 'This Declaration may well become the international Magna Carta of all men everywhere,' she argued, also likening it to the American Bill of Rights.[2] There were forty-eight votes in favour and eight abstentions: from the Soviet bloc, South Africa and Saudi Arabia. But, crucially, not one felt able to vote against it.

The document had survived the arduous process of drafting largely complete. In the words of Harvard's Mary Ann Glendon, 'For the first time in history, the organized community of nations had issued a common declaration of human rights and fundamental freedoms. With its claim to universality, the Declaration marked a new stage in humanity's quest for freedom.'[3] Nor was it the end. In 1966 the General Assembly adopted the International Covenant on Civil and Political Rights and the International Covenant on Economic, Social and Cultural Rights. Both came into force in 1976, giving binding legal meaning to human rights on an international level.

The Declaration was a triumph for Roosevelt and her fellow drafters – she helped give life, purchase, to a big idea centuries in the making. But it was and is a human triumph as well, and a reminder of how big ideas can still impact the world.

*

Think about how to realise a big idea and you might imagine something like the moonshot, a colossal effort on a defined task to realise something new. The Universal Declaration of Human Rights is also a kind of moonshot. Roosevelt and the Universal Declaration mark the moment where the idea of human rights finally found global purchase. It wasn't easy. The framing process was arduous, involving many people from around the world, a layered mosaic of thought, opinion, argument and tradition. Steps forward were frequently rebuffed. Progress was measured in inches as the Soviets and others wore time down on the clock. Roosevelt's name, along with that of Malik, stands tall, but this was a team effort.

But given the ambivalent picture I have presented in Part I, it is important to remember that this idea, like most ideas, was built on centuries of work. This moral moonshot didn't take shape during overheated days of debate in UN committee rooms – it sprawled over decades and centuries. Like most big ideas, the timeline of human rights is long.

The idea of innate and inalienable human rights is in one form or another ancient, perhaps going back to Magna Carta or Cicero or even the Code of Hammurabi in ancient Babylon.[4] But it was during the Enlightenment that the discourse of human rights took recognisably modern form. Thinkers began to argue that reason and moderation could lead to happiness and a better life. Peace, progress – these were, with the right application of analysis, intelligence and moral effort, possible on Earth.[5] The Enlightenment began a shift to more reasonable, curious, cosmopolitan, sceptical stances; as much a shift of emotional disposition and sensibility as the articulation of cold rationality – both of which are important for rights.[6]

Coming in the wake of centuries of bitter religious conflict, human rights were built on appeals to reason and an extension of sympathy – at the time a daring postulation of the existence of a 'natural law' emanating from humans' innate dignity. Thinkers began to argue that rights were inherent, applicable equally to all humans. John Locke's *Second Treatise of Government* from 1690 sketched the framework of basic rights. He argued that people have a natural right to liberty, self-preservation and property, helping to found a new classical liberalism and giving concrete shape to a set of naturally produced rights. Thomas Paine's interventions in America and France – *Common Sense* and the *Rights of Man* – later elaborated a concrete and more developed vision. All the talk of 'Man' did not go unnoticed: French playwright and activist Olympe de Gouges' *Declaration of the Rights of Woman* and Mary Wollstonecraft's *Vindication of the Rights of Woman* made forceful cases of that. Alongside the intellectual history came a parallel story of political development: the British Bill of Rights of 1689; the US Declaration of Independence in 1776 ('We hold these truths to be self-evident, that all men are created equal, that they are endowed by their creator with certain unalienable rights, that among these are life, liberty and the pursuit of happiness') and the American Bill of Rights; the French Revolution and its Declaration of the Rights of Man and Citizen of 1789. These are stepping stones to the modern idea.

As much as we can trace the genesis of modern human rights to the Enlightenment, so limitations are powerfully evident. The unignorable fact of slavery made a mockery of 'natural' human rights. Olympe des Gouges was guillotined in the post-Revolutionary Terror for her beliefs about the rights of women, Paine imprisoned for his efforts. The Terror itself was a reminder of how quickly moral purpose could be undermined. What's more, there was still little sense of rights beyond the limited freedoms of classical liberalism. Only the great social movements of the nineteenth century would deliver a wider conception of human rights that involved aspects like the right to work or the right to an education. Indeed, some scholars argue the Enlightenment idea of rights has been overplayed.[7]

It was a long way from John Locke to 1948. But it was also a long way from 1942 to 1948. As Hitler's blitzkrieg and genocide subsumed Europe, it must have seemed impossible that just a few years later something like the UN would implement a declaration of human rights. And were the General Assembly's vote to have happened just a year later, it is unlikely, with fraught events in China, Korea, the Middle East and Eastern Europe, that the world could have reached such a consensus; it went through in only a brief window of Soviet-American détente.

Nor should the Universal Declaration and the purchase it implies give any sense of a job done. Even today there is no legal or philosophical agreement on human rights, what they are, how they are grounded. Many still argue that, far from being universal, they are impositions from the West. In practice human rights are still, of course, frequently ignored and denied. Unresolved questions of balance linger: between freedom of expression and personal privacy, for example, especially in the context of quickly evolving communications and monitoring technologies. Some ask whether rights should be extended to animals or even plants, or perhaps ecosystems or the planet.

The frontier is ambiguous rather than a clean-cut space; its rollback is the uneven, fragile, still contested work of centuries.

And yet if the message of Part I was that things could be looking better, we shouldn't lose heart. Human rights are evidence enough of that; of how bold suggestions on paper, failed experiments and slow-accumulating undercurrents can build and spring into the open after decades of stasis. Human rights conform to the picture of luck and multiple discovery sketched in Chapter 1, subject to immediate contingencies but also the result of underlying pressures. Big ideas are then the projects of decades and centuries; years go by with seemingly no progress, but then comes a sudden reversal, breakthroughs out of nowhere. Ideas work both on timescales of movements like the Enlightenment, and on knife-edge speeches and quiet, unforeseen demonstrations like those of the Wrights. It is at once a delicate, local process, down to individual pamphlets and speeches, personalities and thinkers, but also one of great earth-changing events like the rise of socialist parties, the American Revolution or the Second World War.

Reasons for pessimism should not be overstated. There is no simple lens of stagnation or acceleration in such a complex process, only an almost infinitely deep concatenation of causes, interlocking on different timescales. If things look good or bad this year or this decade, that is never conclusive. Even if big ideas look uneven, are uneven, a lot may be brewing. Slowdown might mask a build-up of forces; a boom time might reflect the exploitation of discoveries made long ago.

Looking to the future, as we will do in Part II, we are never likely to catch all the contingencies, identify the next Roosevelt or Pasteur, predict where, when and what they'll do, foresee the small but influential tumult of events that will trigger or block ideas in an immediate sense – how the key speech will be received, whether the grant application is approved. But those underlying factors can be identified. You can ask: are we in a world that is more or less likely to give rise to Roosevelts and Pasteurs, to cheer the speech, greenlight the risky application? You couldn't predict 1948 very easily in 1938; but Enlightenment dreamers could still sense the potential and direction of history, the pressures of ideas mixing and igniting as conditions built on the ground, the drift of opinion, the weight of a big thought growing. Ultimately those deeper structural factors were what told, what enabled Roosevelt, Malik and the rest to do their work. It is to those structural factors, both impeding and spurring on ideas, that we now turn.

Part II

BIG IDEAS TOMORROW

5

The Idea Paradox

The Other Nuclear Winter

3 September 1948, Oak Ridge, Tennessee

It was a nervous day. Nuclear power was back. It had ended the Second World War, but its legacy was the mushroom cloud, radiation sickness and the threat of Armageddon. Could it have a better use? There was, hopefully, to be a literal lightbulb moment: a vast reactor, built under the greatest wartime secrecy to produce plutonium from uranium, was now to power a single lightbulb. Oak Ridge laboratory, so valuable to the enrichment of uranium and the final phases of the war, had made the pursuit of nuclear energy a priority for the peace. After three years of frenetic work they had a prototype: an adapted version of their X-10 Graphite Reactor, built at the height of the conflict in just eleven months.

The first controlled nuclear reaction had happened just months before the X-10's debut, on 2 December 1942, led by Enrico Fermi on a squash court at the University of Chicago. Known as Chicago Pile-1, it was a crude stack of graphite and uranium producing a controlled fission reaction, where the nuclei of a heavy element like uranium split into two.

Progress on the X-10 Graphite Reactor had been astonishingly rapid. Just a decade before, in the early 1930s, the ideas behind nuclear power weren't even fully understood. Ernest Rutherford had only discovered the structure of the atom and so launched the nuclear age in physics in 1911. But in the wooded Tennessee hills scientists were bypassing the apocalyptic nightmare of Hiroshima and Nagasaki for what they hoped would bring an age of limitless energy.

The reactor itself had been part of the Manhattan Project, the budget-busting effort to produce an atomic bomb. Absorbing resources equivalent to $200 billion today, the finest minds in physics and engineering worked around the clock to produce a war-ending weapon. The original idea of nuclear fission, the splitting of the atom and energy release behind the bomb, had only come about in the late 1930s, when a group of German émigrés, notably Otto Hahn and Fritz Strassman, began to realise the significance of their work over the previous decade. This in turn built on an extraordinary body of research and imagination in chemistry and physics. It had been a golden age of discovery and theory from the experiments of the Curies to the path-breaking insights of quantum theorists, leading to 6 August 1945 when the B-29 Superfortress Enola Gay dropped 'Little Boy', a 9000 lb uranium-235 bomb, over the city of Hiroshima. Nuclear progress was fast, supercharged by the existential crisis of global war.

On 3 September 1948 the reactor was again switched on. It worked. Nuclear power went live to the grid for the first time on 27 June 1954, in the Soviet city of Obninsk. Two years later, at Calder Hall in England, a young Queen Elizabeth switched on the world's first fully fledged nuclear power plant. From the first experimental and theoretical breakthroughs, it was an intellectual and engineering express train; inventions and ideas advanced at helter-skelter pace, reordering our understanding of the universe and transforming our mastery of it. In the late nineteenth century no one had even heard of nuclear physics; decades later it loomed Colossus-like over the age, threatening extinction and powering fridges alike.

When postwar physicists began to talk of an even better, cleaner, more powerful source of energy – nuclear fusion, imitating the process of the stars – it was naturally assumed this was round the corner; fusion would be with us within a decade, at most thirty years.[1] Not only was fusion safe and without harmful waste products, but its inputs, two isotopes of hydrogen called deuterium and tritium, were plentiful compared to those of fission power. Fusion was an idea whose time had come.

Famously fusion power is still thirty years away; as fusion physicists joke, fusion is always thirty years away. But the challenge – which requires creating plasma, a fourth stage of matter, heated to a temperature hotter than the core of the Sun – is immense. The Manhattan Project was already one of the biggest scientific and energy endeavours of all time. Matching it in peacetime was a huge ask. Unforeseen problems meant the challenges were consistently greater than anyone anticipated.

In fusion, hydrogen is heated to hundreds of millions of degrees. Magnetic fields hold it inside what is called a tokamak, a doughnut-shaped ring. At that temperature the hydrogen nuclei collide and fuse to form helium nuclei. This process releases energy – and presents a fiendishly difficult engineering problem. Increasing the scale of the tokamak, for example, was understood to make fusion more likely. So you build a bigger tokamak. But this creates a welter of snags. And, of course, costs far more money.

Theoretically, fission requires little energy to get started. Fusion requires us to recreate the core of a star on Earth. The idea of fission came off the back of huge leaps in the understanding of nuclear physics. Fusion relies on the manipulation of plasma, a process inadequately understood when research began. Fission had military application in the immediate postwar years as a power plant for submarines, a government 'must-have' that fusion struggled to compete with.

Over the decades came several false flags. The Zero Energy Thermonuclear Assembly in 1950s Britain almost seemed to have cracked it but was a false alarm, as was the hype around cold fusion in 1989 when two University of Utah chemists claimed to have done the seemingly impossible. At every stage, progress threw up new challenges. In Princeton, Oxfordshire or Moscow, it was a case of two steps forward, 1.9 steps back.

Fusion's power record was set in 1997 by an experiment at the Joint European Torus in Oxfordshire, resulting in a reaction whose power output was 60 per cent of that put in. While an international coalition invests billions in a vast new fusion project called ITER, a 5000-ton behemoth, and new startups like the SPARC reactor at MIT are claiming to drastically speed up the process, many scientists remain sceptical of quick progress. Laser compression techniques could proffer a breakthrough, but are still not ready. ITER isn't scheduled to deliver power to the grid until the second half of the twenty-first century. From the beginning ITER's vast scale and genesis as part of an international coalition made progress achingly slow. The whole project almost didn't get off the ground, sunk by bitter inter-country squabbles. Politics still dogs the build.

It turned out that the problems involved were far greater

than expected; fusion needs multiple engineering and scientific breakthroughs whose depth and interplay are, to date, insurmountable. Funding and research requirements are perennially stretched thin. Progress has stalled.

Civilisational advances are predicated on an expanding and secure energy base. Societal collapse is strongly correlated with crises in energy supply. And yet even as the polar ice melts and the deserts expand, the seas overheat and the tropical rainforest burns, a comparatively negligible proportion of global resources are invested in this game-changing technology. Since 1953 the US government has spent around $500 million per year in today's money on fusion: less than a single stealth bomber. Meanwhile global government R&D on energy is around $27 billion a year, less than the US spends on pet food – minuscule in absolute terms.[2] Cat treats and cosmetics are allocated more resources than saving the planet.

In theory Oak Ridge and its counterparts have never been more capable than today, never more blessed with resource or talent – Oak Ridge alone has a budget of $2.4 billion, is the centrepiece of the US Department of Energy's research ecosystem, and hosts the world's most powerful neutron beam and one of its most powerful supercomputers. Yet fusion remains a dream. The great Soviet fusion pioneer, Lev Artsimovich, was asked when it would be ready. 'Fusion will be ready when society needs it,' he said.[3] But this big idea is coming too late.

The story of fusion – and where it goes next – illustrates how ideas can stop pushing back the frontier in great leaps, and start making progress inch by inch, experiment by costly experiment. Ideas might be conceived but then go no further, leaving only unrealised outlines, so many castles in the sky.

The question is why? And will it continue? Those two questions, not just for fusion, but across the frontiers of humankind, form the heart of Part II. In the next two chapters we will explore two broad forces weighing against the future of big ideas. They explain our mixed record at the frontier, why ideas are getting harder to conceive and execute, and steer the future onto a slowing and troubled trajectory.

First are innate problems in the nature of ideas themselves. Nuclear fission does not make fusion a done deal. In fact, solving one problem leaves only the harder problems to tackle. Unsolved problems are unsolved for a reason. Sometimes progress in one area supercharges it across domains. At other times you hit a local maximum. Having more ideas doesn't mean automatically speeding up the rate at which big ideas are generated; not if the challenges that go with them are correspondingly scaled up as well. All the while, as you ascend through the gears of discovery or invention, you approach certain limits. Challenges of this kind, woven into the nature of ideas themselves, are discussed in this chapter.

Second, social context hasn't helped fusion. Although climate change may ultimately offer an existential challenge dwarfing even that of the Second World War, to date there has been nothing like the urgency of the Manhattan Project. Instead society prioritises short-term expenditure. Funding inevitably breeds political difficulties, whether interstate rivalries or simple commercial nationalism. Smaller-scale ventures get caught in a trap of overpromising and underdelivering. The regulatory burden is remorseless. Financial, managerial and political roadblocks are legion not just for fusion, but across the board. These societal barriers will be explored in the next chapter.

How we respond will define us. Luckily there are also reasons for optimism. In the following chapters I look at forces that stand ready to catalyse new big ideas at a pace and scale never before witnessed. Caught in the middle of this delicate interplay is the future.

Eating the Low-Hanging Fruit

For most of history, it was by no means obvious what was an irreducible material – what we now call a chemical element. But investigators discovered the building blocks of the universe and, in so doing, built an extraordinary foundational account of chemistry.

This history displays an uneven gradient of progress. Some elements, like gold, copper or iron, had been known for centuries. Early experimenters developed an understanding of elements like carbon and sulphur. From there, though, an infrastructure of techniques and tools, knowledge sharing and accumulation was required for exploration to keep going. Nonetheless, as the picture began to fill out individuals were still capable of making a huge impact. In the late eighteenth century, the British scientist Sir Humphrey Davy alone predicted the existence of elements like potassium, sodium and calcium, and was then able to isolate them. Around the same time the discovery of fundamentals of chemistry like hydrogen, oxygen and nitrogen altered the chemical lexicon forever.

It was the more plentiful elements that were usually found first, relatively simple experiments sufficing to confirm their existence. Then it got harder. Progress slowed, before a breakthrough technology or insight opened a door and the process was repeated. For example, the rate of discovery dipped in the middle of the nineteenth century before accelerating at its end. When Mendeleev first announced the periodic table to the Russian Chemical Society in 1869, it included sixty-three elements. Soon after, new techniques enabled the discovery of noble gases like argon and neon. Then came radioactive elements like polonium and radium. But this was all much more difficult than stumbling across iron or even isolating oxygen.

Fitting the wider pattern of diminishing returns and rational pessimism, the number of elements discovered per chemistry paper published then declined right up to the present.

Discoveries continue, but you need enormous equipment and extremely rare source materials to find these 'superheavy' elements. They are elusive and unstable, decaying in moments. Indeed, one Russian lab has built a $60 million experiment just to try and find the elements numbered 119 and 120.[4] Compare the journey of discovery for two elements a few centuries apart:

> element 117 was discovered by an international collaboration who got an unstable isotope of berkelium from the single accelerator in Tennessee capable of synthesizing it, shipped it to a nuclear reactor in Russia where it was attached to a titanium film, brought it to a particle accelerator in a different Russian city where it was bombarded with a custom-made exotic isotope of calcium, sent the resulting data to a global team of theorists, and eventually found a signature indicating that element 117 had existed for a few milliseconds. Meanwhile, the first modern element discovery, that of phosphorous in the 1670s, came from a guy looking at his own piss.[5]

Nobody knows how many elements potentially exist, but it's clear that when it comes to the periodic table, the low-hanging fruit was naturally picked first. Regardless of where the table ends, the direction is towards ever more difficult discovery. There will be no more Humphry Davys.

That we pick the low-hanging fruit first is a truism. But its impact is enormous and almost wholly overlooked in policy, business and intellectual life, even among many who engage with the Great Stagnation Debate. Not only do we pick them first, but the world is unidirectional; save for extreme calamities like the collapse of the Roman Empire, once a discovery has been made, a technique implemented or a thing invented, it stays made, implemented or invented. That is to say, big ideas happen and are achieved once.[6] They might have long histories, numerous tweaks and subsequent revisions, but this doesn't hide the fact that once a big idea has been conceived, executed and found purchase, doing the same cannot rival it. The frontier has already moved. This fact alone explains much of the present's mixed track record. And it is part of why the future contains such daunting challenges.

Although the theory is controversial, more and more scientists, economists and thinkers argue that the low-hanging fruit problem is real. Barring the intervention of transformational tools, discoveries or circumstances (the grand questions, all of which will be discussed), success at generating and delivering big ideas raises the bar for further big ideas. All things being equal, realising, delivering, instantiating, imagining the significant ideas of the next one hundred or two hundred years will be – objectively – more difficult than those of the past. This emphatically does not mean there are no more low-hanging fruits. The argument is not that they don't exist, only that in future they will be fewer and trickier to obtain.

Economists use this concept to explain slowing growth and the introduction of new technology.[7] In the 'special century' already discussed, the average human life went from rudimentary – most people living in quasi-medieval conditions – to, roughly, barring some digital bells and whistles, an existence like ours today. As we know, everything from telephones to radio, fridges to central heating, motor cars to electric light was introduced during those years. Despite the average family having vastly more wealth than in the 1950s, these improvements are effectively one-time events, and so that increased wealth doesn't translate into a kind of never-ending experiential escalator.[8] Ubiquitous light or a warm home and the big ideas behind them have happened: no other home lighting or heating idea will have equivalent impact. You can try re-inventing the wheel, but it will still be a wheel.

An equivalent process can be seen in the arts. There can be only one Andy Warhol, Doris Lessing, Elvis Presley or Zaha Hadid, in that repeating their techniques never has the same impact. After the breakthrough is made, the idea established, the fruit eaten, the frontier shifted, it may cast a long shadow, but again it has happened, the world is changed. Repainting, rewriting, redesigning, reperforming is not equivalent. Once an aesthetic space has been opened, it's open. Play in it all you want, but the key moment has already occurred. The same applies to scientific discovery; Copernicus might be revised, corrected and superseded, but (if we can set aside the Greeks for a second) the basic move – shifting our universe – has been made.[9] I can paint like Picasso, suggest that an Internet search engine would make a clever business, or isolate oxygen but no one would accuse me of originality. This doesn't mean that new things won't come up in the future; it means we should expect a steeper bar, because 'obvious' avenues at the frontier are already taken.

Charles I. Jones talks about 'fishing out' as a possible mechanism by which the generation of ideas gets harder: in the great pond of ideas, the fat and lazy fish have all been caught. That is, ideas get harder as ideas are thought. Are we all fished out? While I don't think we are, Jones is right to say, given the breakthrough problem and diminishing returns, that easy pickings are scarce and getting scarcer.

Ideas lie on a ladder of ascending difficulty or obscurity. The tools, methods and paradigms of the past cannot always make the climb. Scaling it rather requires external, frequently unforeseen breakthroughs, or more resources, or old-fashioned luck. Often all three. We therefore rely on circumstances being equal to the challenge, but there is no necessary reason they ever should be. It's not that problems in the past were easy, the fruit lazily waiting; just that those strides forward moved us on to new peaks. Now they surround us.

Archaeology presents an interesting analogy: techniques like geophysical surveys, sophisticated dating technologies, drone mapping, autonomous submersibles, big data analysis of satellite imagery, lidar scans and 3D imaging amplify the discipline. Archaeologists can peer deep beneath the earth or within structures without having to lift so much as a trowel. They have more knowledge, more ability to collaborate and share data, expertise and tools. The capacity for making significant archaeological discoveries has never been greater and, indeed, impressive results have been delivered in recent years – from a complete Greek galleon to some of the earliest known paintings, sixty thousand hidden Mayan structures, mummy workshops, a palace of Ramses II and evidence of human migrations and ancestors earlier than any previously suspected.[10]

But the impact doesn't compare with the early days. There was a golden window for breakthrough archaeological finds, like Arthur Evans' coming across Minoan civilisation at Knossos, the great Heinrich Schliemann's uncovering of Troy, Yale-affiliated Hiram Bingham's adventurous rediscovery of Machu Picchu, the opening by Howard Carter and Lord Carnarvon of Tutankhamun's tomb in Luxor's Valley of Kings, Leonard Woolley's excavations in Iraq, and the discovery of Sumer and Ur and Uruk, cities older than had been known. Not to mention the caves at Lascaux, the Dead Sea Scrolls . . .

This is no skin off the nose of contemporary archaeologists. It's just that back then the low-hanging fruit were bigger and riper, and no matter how much we throw at it, we're unlikely to be able to repeat such stunning finds. Significant discoveries become less likely over time, even as the capacity to make them grows. We won't find Atlantis because, truthfully, if it existed they'd already be offering guided tours and fridge magnets.[11]

Archaeology isn't a universally applicable illustration, in that there are a finite number of objects or sites to be found. It does however illustrate how when a field or technology or business is new, its growth accelerates fast, progress comes in leaps and bounds. But this cannot go on forever. Once the low-hanging fruit have been plucked what remains are more obscure concepts, knottier problems, mature markets, dustier corners in the attic of invention.

The nature of ideas means we must ascend rungs of difficulty and obscurity – and will keep on doing so. Despite being frequently dismissed, the low-hanging fruit phenomenon is ingrained. Science, technology, human civilisation – unlike archaeology, these may be boundless territories. But we reach – and have reached – the more navigable destinations first. The corollary is clear. All things being equal, future destinations will be harder.

The Burden of Knowledge

Few people in history have known as much as the physician Thomas Young – a man 'eminent in almost every department of human learning', as his tomb in Westminster Abbey has it.[12] Born in Somerset in 1773, Young always showed signs of extraordinary polymathy. Trained as a doctor, he practised medicine throughout his varied life, but that was only the beginning. He was also a lecturer at the Royal Institution aged just thirty, covering the entire range of science, while he served as the foreign secretary of the Royal Society and the secretary to the Board of Longitude among a host of other appointments, and dabbled in the industry of life insurance.

As an independent scholar, Young proved Newton wrong. By demonstrating the interference of light he proved that it was a wave, not a stream of 'corpuscles' as Newton had believed. Young showed how the eye focused and registered colour, laying out the understanding of vision. And yet it was this same Thomas Young who helped decipher the Rosetta Stone, so unravelling the mystery of Egyptian hieroglyphics even if he was beaten to the punch by his (more specialised) French rival, Jean-François Champollion. No doubt Young's prodigious mathematical ability and knowledge of fourteen languages from Aramaic to Amharic helped.

He was the first to define energy in the modern sense and furthered the physics of surface tension; he systematically compared the grammar and the vocabulary of 400 languages, introducing the name (if not the concept) 'Indo-European' to describe that family of languages; he produced systems for tuning musical instruments and administering correct doses of drugs. His name is honoured in Young's modulus, the Young–Laplace equation, the Young–Dupré equation and the Young Rule. Asked to contribute to the *Encyclopædia Britannica*, and on the condition of anonymity, he offered to write on 'Alphabet, Annuities, Attraction, Capillary Action, Cohesion, Colour, Dew, Egypt, Eye, Focus Friction, Halo, Hieroglyphic, Hydraulics, Motion, Resistance, Ship, Sound, Strength, Tides, Waves and "anything of a medical nature".'

Perhaps, as his biography's title has it, he really was the last person to know everything. Young stands at the head of a long line of super-polymaths, from Lucretius to Leonardo to Leibniz: the *uomo universale* or Renaissance man. And yet of course, Young did not know everything. The extent of knowledge was already far beyond him, let alone Lucretius and Leonardo.

Young was probably one of the last people to span so wide an arc of enquiry at the frontier. Knowledge was already too unmanageable – an ancient plea but a powerful one nonetheless. The Renaissance man was dying by the time Young tried and failed to know everything; today, they are definitively dead.

As knowledge advances, there is always more to know. Greater specialisation and longer training times result. Benjamin F. Jones calls this the 'burden of knowledge' effect, whereby the sheer quantity and balkanisation of knowledge becomes an obstacle to original thought. Its ramifications are huge and underappreciated, and we should expect it to intensify in future. To gain a sense of this burden, think of libraries: in the seventeenth century John Harvard was given the naming rights to his eponymous university by donating some 320 books. Today the Library of Congress alone holds 38 million volumes.[13] Harvard itself has 20 million. That exponential increase in the number of books is the burden of knowledge.

Breakthroughs happen at the frontier of knowledge. But the distance to that frontier is always getting further away. All of us start learning at square one, with nothing. We must travel to the frontier, and that process – of studying, mastering, memorising and practising – gets longer as knowledge accumulates. Jones has quantified this trend, modelling the 'burden of knowledge' via a corpus of 55,000 patent inventors drawn from the United States Patent and Trademark Office from 1975 to 1999. The numbers of people behind a patent and their age at the grant of first patent are both rising. That is, it takes longer to create a patent, and even then, it needs more hands. There was an average of 1.73 inventors per patent in 1975, but by 1999 it was 2.33, a 35 per cent increase over the period, and the researchers behind this figure argue it is an extremely conservative estimate.[14]

Papers published in the *Proceedings of the National Academy of Sciences* have more than double the number of authors since the 1990s, and none of this is to mention what we've already seen, the rise of mega-authored papers.[15] While this change is perhaps unsurprising in the resource-hungry sciences, the pattern of growing teams holds in traditionally solo areas: in economics, mathematics, the arts and humanities the most important work is now team derived, just as creative productions involve more people over time despite improvements in technology.[16] The more knowledge already exists, the larger the team required to generate new knowledge or invention. And the larger the 'citation tree' – the number of earlier patents cited in the creation of a new patent – the larger the team.

When I talk about things being 'harder' or more 'difficult', this illustrates what I mean; that mechanisms like the burden of knowledge imply that it takes more people to have an idea now than in the past; and having an idea in the future will need more people than today. Larger teams bring with them greater coordination costs and interpersonal and institutional frictions and constraints. An analysis of more than 65 million papers, patents and software products published in *Nature* showed that while large teams were good at developing existing ideas, 'smaller teams have tended to disrupt science and technology with new ideas and opportunities'.[17] The higher-impact the work, the more this was true. Both are important, but the implications for radical new departures – big ideas – are troubling.

Specialisation has also been increasing for centuries. More recently, measured by inventors filing patents in different fields (controlled for various influencing factors), the data show a 6 per cent increase in inventor specialisation every decade. Just one example: in the fifty years between 1960 and 2010, the number of unique pairs of terms in articles categorised by the US National Library of Medicine grew from 100 to 100,000.[18] As specialisation grows, knowledge production turns towards minutiae, at times towards the trivial. In other words, the very structure and trajectory of knowledge, the great expansion of the frontier itself, adds to the challenge of coming up with a big new idea in particular.

A corollary is that as researchers become mired in tiny niches they are less likely to switch fields. They remain stuck in narrow disciplines and worldviews; fresh perspectives and the chance of cross-pollination get lost. As we will see in the next chapter, our social and academic structures are complicit. As it recedes, the frontier gets narrowed down to nested sub-sub-sub-specialities, funnelling people into ever smaller portions of enquiry. In the 1950s a biochemist or a linguist might look at DNA as a whole or create a theory of language; today they would look only at a minute fraction of either. Occasionally people break out – but all too often, the best they can do is to create a sub-sub-sub-sub-speciality. This extreme narrowing effectively precludes broad thinking at the frontier, without question one of the most powerful dampeners on the future of specifically big ideas.

It also dovetails with the increased length of researcher training periods. Jones shows how the age at which scientists or inventors make their first invention has remorselessly risen.[19] People are older than ever when they do their most significant work; it takes them longer to see through their most meaningful ideas.

The age at which doctorates are awarded has consistently risen since the 1960s and there has been a marked increase in the number of postdoctoral positions people occupy before leading their own research; between them they often take well over ten years. Jones finds that in the early twentieth century future Nobel winners started independent research at twenty-three. The average age was thirty-one by its end.[20] The average age of recipients of the prize, and when they did their winning research, has risen remorselessly across physiology, chemistry and physics.[21] Today you apparently need to know much more to make a discovery. An ageing effect hence shortens the most significant idea-producing years of researchers' lives.[22] A 30 per cent decline in innovation potential can be explained by this effect. Nor does this imply that people make up the difference with greater productivity when they hit middle age. In fact all that extra work doesn't increase people's potential in the long run.

Does age matter? Newton was twenty-three when he made many discoveries; Steve Jobs was twenty-one when he co-founded Apple, while Mark Zuckerberg was nineteen when he founded Facebook. Mozart, Évariste Galois, Mary Shelley, Alexander the Great ... Plenty of people achieved a lot in their early twenties or younger. Moreover, when knowledge undergoes a revolution, the average age of significant contributors falls. The stock of knowledge in post-revolutionary fields is lower; there are fewer barriers to entry; and younger people, not institutionalised in the old ways, are free to rampage across fresh terrain.

This has happened with quantum mechanics, Expressionist art and in Silicon Valley.[23] As quantum physics rewrote the discipline, so there was a lowering of contributors' age: Werner Heisenberg was twenty-five when he made major discoveries and Bohr, Wolfgang Pauli, Paul Dirac and Fermi all made important discoveries in quantum mechanics or sub-atomic physics under the age of thirty. Kuhnian revolutions like this create windows of opportunity for young researchers unschooled in the minutiae of the old paradigm. Dirac and Einstein may have been twenty-six when they started to produce their best work, but it is unlikely a 26-year-old physicist would do the same today. Young people simply have less opportunity to work at the frontier.

A study of near misses in breakthroughs suggests, moreover, that the primary reason behind them is 'paradigm rigidity'; people fail to make the necessary connections and insights because their investment in the current system leads them to miss anomalies, reject solutions and overlook connections.[24] Today, rather than the young and iconoclastic, it is the old, those more likely to be cautious and with something to protect, who dominate.[25] As the stock of researchers ages, it gets more deeply invested in current paradigms and we can tentatively expect fewer revolutions per researcher; more missed breakthroughs.

The impact of ageing is felt not only in research, but also on the economy: much of what we will see in the next chapter can be explained by the general ageing effect of society. Nor is this mere prejudice or supposition: analysis of 20 million research papers found that younger researchers were more likely to try out new ideas, while older researchers were more conservative and less exploratory in their strategies.[26] In accordance with Max Planck's dictum that science advances one funeral at a time, it seems that the death of a superstar researcher does indeed open a field to new outsider perspectives which spark progress.[27]

In the future it takes ever longer to master a field. An individual's knowledge of the distant frontier comprises a smaller and smaller myopic slice, further removed from other disciplines. Big ideas not only require more people but those people are older, operating on narrower terrain, encountering more friction. Ultimately this picture is built on what Benjamin F. Jones calls 'accumulations of knowledge within fields'. In other words, as time goes on specialisation and workload increase: 'An intriguing side effect of innovation is the possibility that new ideas impose an increasing educational burden on future innovators.'[28]

The significance of the burden of knowledge effect is twofold. First it largely explains the diminishing returns seen across fields: the mechanisms behind Eroom or Jones et al. The second is that, as we saw earlier, knowledge production is still increasing on robust trendlines. We can therefore predict with confidence that the burden will keep rising. And if there are big ideas to be found between fields, then the weight of specialisation will press doubly hard. The more narrowly research is forced to tunnel, the harder it is to connect those tunnels.

Benjamin Jones argues that 'fishing out' is the wrong image.[29] It's rather, to adapt the metaphor, that the fish are harder to find, deeper in the lake, better at not taking the bait; you need more people trying to catch them with longer, more unwieldy and specialised rods. The more ideas you 'catch', the more new apparatus you need in future. Absent a totally original technique or fishing rod, it's inevitable that the catch will thin as you progress. For me this is the right take. It implies that, provided you can and do build bigger or better rods, fish are there to be landed. But it acknowledges that ideas exist on a gradient. That ideas always needed to be caught rather than just appearing, they were always hard to land.

Another way of understanding the growth in team size and the increase in specialisation is to look at the flipside: everywhere fields exhibit greater complexity. Again this doesn't preclude big surprising ideas, but it introduces further layers of difficulty.

Think of both physics and policy. In physics, one story of the twentieth century was about the descent to ever more fundamental particles and theories. Going through the layers meant hitting ever more inaccessible areas; the task of getting there grew more complex. But equally, crafting new policy has also become more complex; additional knowledge, considerations, interest groups and existing regulation mean that, with every passing year, the burden of proof and the raw complexity of designing and delivering new policy initiatives only increases. As complexity increases, so does the difficulty of making a breakthrough.

Already our societies face an 'ingenuity gap' which renders us incapable of dealing with the complexity of events.[30] A system's complexity is rarely understood by any individual, a phenomenon shown to be behind industrial accidents like Three Mile Island or the BP oil spill.[31] A single Boeing 747-400 contains over six million parts and 171 miles of wiring.[32] Software now typically runs to millions of lines of code and runs on devices engineered at nano scales. In the twenty years since its creation in 1990, the program Photoshop's codebase grew by over forty times.[33] This is typical of a ballooning underlying complexity in software. What's more, evolutionary software self-iterates, automating that process towards incomprehensibility.

Over time layers accrete in the system. Various exceptions and edge cases get factored in, whether relating to legal code or software use, piling on more complexity. No one can get the total view or understand all the elements and how they interact. And every increase in complexity even on sub-components, whether of an urban infrastructure, a market, a technology or a legal system, necessitates more specialisation, more distance to the frontier. These interlocking systems negate big, bold interventions and demand compromise and careful iteration. They have too many intricate working parts, whether knowledge of a discipline like biochemistry or the regulatory framework around launching a new drug.

A more complex world changes the way ideas and fields

emerge. Eureka moments recede further from reality. The study of complexity (commonly understood to describe non-linear, aperiodic systems that are acutely sensitive to initial conditions) is itself a good example.[34] Also known as 'chaos', it never really formed a discipline unto itself. Instead it's typical of a transformed genesis of big ideas: at one level it elegantly unifies, bridging dispersed phenomena. Yet building a complete picture of the subject is beyond the competencies of any individual or even discipline.

Chaos theory may have originated with the use of computers to model the weather, but it ended up programming those computers, saving endangered species and forecasting global economic flows. It encompassed mathematics, meteorology, population ecology, epidemiology, physics and economics, underwritten by advances in computing and computer science and applied to topics as various as the organs of the human body, atmospheric storms and a beam of particles. Names like Edward Lorenz, Mary Cartright, Benoit Mandelbrot and Mitchell Feigenbaum were all closely connected, but it didn't have one great progenitor in the manner of a Darwin or Pasteur. The study of complexity is itself, perhaps unsurprisingly, complex, transdisciplinary, composed of different foci and sub-ideas, while nonetheless adding up to something big and revolutionary. It doesn't even have an agreed name. It isn't a discipline as such; in this, it's typical of how big ideas will form in future.

While the borderlands between disciplines are ripe territory for major progress (if there are low-hanging fruit, that's where we'll find them), it not only adds to the burden of knowledge, but creates issues of coordination and clashing worldviews.

For example, genetic science now has to incorporate cutting-edge research from not only biology, but also chemistry, physics, psychology, economics and sociology, with their differing methodologies and mindsets. *Science* called epigenetics 'the greatest paradigm shift in recent history'.[35] Long considered scientific heresy, the supposedly discredited legacy of Jean-Baptiste Lamarck, epigenetics contends that contextual conditions impact heredity. Via complex epigenetic mechanisms, environmental factors *may* impact the expression of genes.

The molecular biology of genetics was already complex, and repeatedly turned out to be more so than even its most far-sighted proponents predicted. For some time it was believed that a single gene might be responsible for a single trait, before it was shown that traits formed from the complex interplay of many genes. Epigenetics takes this to a new level. Psychological factors, the trajectory of a life, even sociological or historical factors, come into play. Although the matter is keenly debated, many scholars now believe that both neo-Darwinian and neo-Lamarckian mechanisms help drive evolution. The range of factors that need to be taken into account of any understanding is increased.[36] It is one instance of a more general rule: simple(r) answers, simple(r) mechanisms, are ever less likely.

Moreover, we can be confident this increasing complexity works exactly like the burden of knowledge: it increases the acquisition costs for knowledge and makes reaching the frontier more difficult. Indeed, so powerful are the constraints of complexity that the researcher Alex Mesoudi believes it may put an upper limit on our cultural evolution.[37] Essentially we could arrive at a point where knowledge and culture are so complex, we are completely bogged down in trying to process and understand what is already there. Accumulations don't always work as a springboard; sometimes their weight holds us back. Compounding the problem, evidence suggests that interdisciplinary work, needed to grapple with the complex questions at the frontier, has consistently lower success in being funded.[38]

Scanning the horizon, we can be sure our frontiers will keep receding, paper by paper, patent by patent. More time will be spent in education; teams will get bigger; problems, systems, technologies, will complexify beyond comprehension or intervention; fields will merge, split and arise unpredictably. Every answer prompts further questions; every move forward unlocks more complex mazes ahead.

To a mariner standing on the shores of Portugal in the late fifteenth century, the oceans must have seemed endless. And yet now, contemplating the prospect of space travel, we've levelled up. It's not to dismiss the courage and ingenuity of those explorers to describe navigating the galaxy as a steeper challenge.

Space travellers know what to expect and can maintain contact with a base. Even the most risk-hungry astronaut of today cannot outdo the explorers of old in buccaneering chutzpah. Nonetheless, the scale of the problem is palpably different. Our closest star and planetary system is over 25 trillion miles away. A ship travelling at a million miles an hour would still take three thousand years to reach Alpha Centauri and involve engineering, biological and psychological challenges whose scope are far beyond anything ever attempted.[39] The leap from a goods-carrying caravel to a Boeing 747 doesn't really do it justice. Transcontinental travel, before modern technology, was immensely tough. It was also a low-hanging fruit. This is where we find ourselves in the twenty-first century. As the challenge of big ideas increases in scale, there is no guarantee that we have the will or means to scale up our efforts to meet it.

Knowledge creates its own problems. We have more powerful understanding, tools and capabilities; yet, inevitably, more intractable problems to go with them. Success is a burden.

Saturation at the Limit

A group of thinkers called mysterians argue that there are hard limits to the brain and thus to our intelligence. Some questions are therefore permanently out of reach. Just as goldfish aren't equipped to understand the germ theory of disease, Maxwell's equations or the novels of James Joyce, perhaps we aren't equipped to solve ancient and apparently intractable philosophical conundrums like the mind/body problem or the fundamentals of the universe. After a point, we may neither be able to find new theories, nor even if we get close, really understand how they work. Some questions are 'cognitively closed'.

Richard Feynman said that 'nobody really understands quantum mechanics.'[40] What if that is the norm in future? If we scrape against the ceiling of our limitations? Although the notion is controversial, if there are limits to what the brain can know or do, we are surely getting closer to those limits. Similarly, given the complexity of our built systems or the detail of found systems, it's not hard to imagine that, should such limits exist, they are getting closer.

The human brain is famously the most complex known object in the universe. But it is also just 1.4 kilos of organic matter evolved over millions of years, optimised for the distant savannah. *Homo sapiens*' cognitive capabilities are bounded by our brains, which are themselves rooted in our deep past. There is no necessary reason why we should be able to produce new ideas ad infinitum or understand massive complexity.

We can only hold about seven things in our minds at once. On some aspects of intelligence, we perform worse than chimpanzees.[41] Our memories are weak – we can memorise and ingest information only very slowly compared to machines.

They can download and index an entire encyclopedia in seconds, but it would take us months or years. Our total long-term memory is something around a gigabyte. Our neural networks are over a million times slower than computers, even if our parallel processing capability is, for now, awesomely more.[42]

We have yawning perceptual gaps and limited senses. Experiences too, such as varieties of aesthetic experience, are closed to us. Unlike most creatures, we are primarily audio-visual, but even here our senses are ineffectual. We can detect only a limited band of the electromagnetic spectrum. Our olfactory powers are shockingly weak compared with many animals – an 'odourscape' of communicative pheromones, alive in every ecosystem, is dead to us. Compared to some organisms we have limited experience of sound, temperature and electric fields.[43] We rely on our perceptions, yet usually don't reflect on how limited they are. It is uncontroversial that we have experiential, sensory limits and very possible that we have cognitive or epistemic limits as well.

If we have eaten the low-hanging fruit and arrived at the hard problems, while all the time adding a burden of knowledge and complexity, there is a further aspect to consider: that there may be limits to what ideas are possible. And further: over time it becomes more likely, not less, that such limits will be hit.

It could be then that at the reaches of cosmology or particle physics, we are approaching the edges of our capability. Many disagree, but it can't be discounted. Yes, we use mental prosthetics – from a sheet of paper to supercomputers – to help. But ultimately, conceptually, if 'we' are said to have an idea, we need a concrete conceptual grasp of it. And we can never be sure of the questions we are incapable of asking. I share the view of Astronomer Royal Martin Rees, who although confident in the power of human inquiry argues, 'We should nonetheless be open-minded about the possibility that despite all our efforts, some fundamental truths about nature could be too complex for unaided human brains to fully grasp.'[44]

Eventually we will grapple not just with our own limitations, but potentially with the limits of the physical world itself. If anything this is even more controversial than the idea that there are limits to our own intelligence. Again though, *if* there is a ground state to things, a final layer of explanation, we can always get nearer. And the more we learn, the more there do appear to be limits to knowledge: Heisenberg's uncertainty principle, Gödel's incompleteness theorem, the philosophical paradox of unknowability, questions of whether anyone could answer questions of infinity around the size or origins of the universe, limits on the total information processing capability of that universe.[45] If there are insoluble questions, we will arrive at those questions in time. Vannevar Bush called science an 'infinite frontier', but as John Horgan points out, this wasn't meant literally.[46] On this point I am agnostic, and many if not most scientists believe we either are very far away or there is no such limit, that the frontier really is endless. We cannot know either way at this stage.

But if there is a technological limit, in say the storage of energy, or the speed of travel, or the potential age of a human being, or the computational capacity of a given portion of matter, we will also get nearer. Perhaps the reason we don't have a warp drive or a time machine or a simulation of the universe is that they defy possibility and we will never have them. In some ways limits may offer comfort, in that we know we are far, say, from the limits of computation – we know for sure there is further to go. But we are heading in one direction. Underneath everything a set of physical constraints bound the frontier, whether scientific hypotheses, or suggestions for new technologies, or cultural experiences. If we hit those limits, it could happen in the next decades. And eventually our ideas, or more likely those of our distant descendants, might reach a hard stop at the edges of existence.

Similarly, although we don't know the limits to conceivable human age or the speed of economic growth, it is unlikely they could ever scale exponentially, as research inputs like resource or the number of researchers have. There are practical limits that make diminishing returns almost inevitable – even if we increased tenfold the resources working to, say, boost human lifespan, we would not expect a tenfold increase in the length of human life.[47] If we had ten times the number of economists and entrepreneurs working on ways to boost economic growth, we would not expect it to occur ten times faster than it is today. We have many more philosophers than the ancient Greeks and many more playwrights than Elizabethan England, but it's not clear we have many more people whose impact at the frontiers is commensurate with Plato or Shakespeare. Rather the picture of diminishing returns painted in previous chapters could be inevitable, baked into the very structure of the fields concerned; we should not expect exponentially rising returns on our exponentially rising inputs, but constant returns. Having to put more and more in just to nudge back the frontier at the same pace isn't sign of a crisis, but part of reality we need to accept.

There's a different but more immediately telling sense of saturation. We have already seen how economics and scientometrics quantifies ideational trends, looking at impact, measuring citations and usage figures, analysing awards patterns, deploying sophisticated statistical techniques. Everyone from these quantitative backgrounds immediately asks me how I measure what is big about a big idea.

But this misses how the 'bigness' of a big idea, even portions of the frontier itself, have an inherently subjective quality. You can never entirely capture it other than by qualitative means. Herein lies a problem. Once we hit a degree of expectation in our view of the universe, in our basic living standards, of a new artwork or business, shifting the dial becomes harder. What would once have been big feels small; the nature of what constitutes a big idea changes.

Consider these examples. We live in a universe of hundreds of billions of galaxies each composed of hundreds of billions of stars. And theories of the multiverse suggest there could be hundreds of billions of universes – at least. One estimate from Oxford scientists puts the size of the universe at 7 trillion light years across, and this could be conservative. Humans are assemblages of trillions of eukaryotic cells. Contained within them are a further 100 trillion microbes supported by that superstructure.[48] Categories like space, time and causality all work counterintuitively. None of this particularly bothers us. Save for the occasional flash of awe, we soon go back to our emails, the washing, tomorrow's agenda. The most extraordinary facts, technologies and insights are rendered humdrum by familiarity.

Wonder does not increase linearly in scale. The leap from a galaxy of a huge number of stars, to a universe of a huge number of galaxies, to a multiverse of a huge number of, well, absolutely everything does not induce commensurately larger gasps. People talk of an infinite capacity for wonder. Maybe not. Next time you're scrolling through Facebook while vaguely watching that mind-blowing nature or physics documentary, check on your infinite capacity for wonder. Eventually it gets saturated.

The notion of the hedonic treadmill suggests that our capacity for taking pleasure can be depleted. We adapt to happy circumstances, and what once gave us surges of joy – perhaps a new plaything or partner or promotion – quickly becomes normal. Something similar has occurred with ideas.[49] What would once blow minds, change the world, crowd the streets, feels much smaller, elicits little more than a shrug. Measuring the cosmos; destabilising the fixed categories of experience: it is hard to compete with these things and they set a benchmark of revolution that is difficult to match.

What shocks us today? Once the world was incensed by a Copernicus, Darwin or Freud. The speed of travel, on trains or planes, stunned our forebears. Artworks from Duchamp's *Fountain* to *Tristram Shandy* scandalised their times. Very little shocks us now. The dangerously transgressive is now just another gesture in a crowded market. When upending assumptions and scandalising the audience is par, the room for upending assumptions and scandalising is diminished.

In other words, big ideas in the future won't feel as big as they did in the past. After a point, change, however grandiose, is subjectively marginal: from an infinite universe to an infinite multiverse, or from house to drum and bass, or from a television to a huge flashy high-definition television, or from representative sculpture to abstract sculpture to found objects. Impact reduces. If you spent life struggling with a horse and cart over mud-bathed dirt tracks or walking miles to get water, the transition to paved roads, a car and running water is almost inconceivably large. For us, moving from comfortable petrol-powered cars with cruise control to electric and autonomous vehicles, the shift is, while significant, smaller. We shouldn't be particularly surprised by this as nothing experiential can keep accelerating or escalating forever. In the words of the Oxford geographer Danny Dorling: 'Sex, drugs, rock'n'roll, schools, jobs, homes, health, beliefs, views, experiences and travel – they cannot always be more different for the next generation than the last.'[50]

Science fiction writers have conjured worlds and

technologies at the edge of what could ever be possible, and we've grown up comfortable in their vast imaginations. Faster-than-light travel? Normal! Teleportation? Standard! *Subjectively* speaking, having a breakthrough idea is harder because we have come to expect them; we have pushed the boundaries so far, the idea of boundaries to break through has become redundant. If you grow up expecting teleportation, anything less is a disappointment. This has an economic parallel: as the economy grows and diversifies, so a jump in any given field has a smaller impact. In the nineteenth century a new agricultural technique or manufacturing technology could have a dramatic impact on an entire national economy. But today even the largest sectors and industries form just a small portion of the whole. Commensurately huge advances in, say, metallurgy, just don't have the same impact, and the sprawl and complexity of the economy means even the impact of general purpose technologies is diluted. As Anton Howes puts it:

> Innovation has, in a sense, been the victim of its own success. By creating ever more products, sprouting new industries, and diversifying them into myriad specialisms, we have shrunk the impact that any single improvement can have. When cotton was king, a handful of inventors might hope to affect the entire national textile industry in some way. Nowadays, there's not a chance.[51]

A similar mechanism from pharmaceutical research is called the 'Better than the Beatles problem'. Because drugs have only a limited period of exclusivity, afterwards they are freely available.[52] How do you constantly produce better drugs than perfectly good out-of-patent alternatives? How can you constantly be better than what is already game-changing?

You might equal it, which is fine; but if your business model requires something unambiguously better, carrying on as you did in the past isn't good enough.

For several hundred years we've been on a great adventure of thought. But that adventure is no longer moving forwards but stuck in an endless self-aware loop. In Chapter 4 we saw how various schools of thought, from Wittgensteinian language games to deconstruction, behaviourism to behavioural economics, told us that objective categories like truth and beauty were nonsensical. That there was nothing universal and everything was relative; there was no external and absolute standard by which anything could be judged. That such fundamentals as nature and the self were flawed social constructions. Art could be whatever you wanted it to be. The idea of any kind of totalising or 'grand' narrative was ripped apart, whether that was a given social order like communism or the power of technology to improve lives. Everything was theoretically equal and, in practice, equally available for use in the new supermarket of concepts.

Moreover an ironic detachment was the order of the day. By the end of the twentieth century you couldn't take anything too seriously – especially ideas. Being surprised and thrilled, realising dreams of the future, is not seen as bracing or even dangerous, as it was for most of history; it is seen as embarrassing or naive. Big ideas became passé, grand statements gauche.

The question is where one goes from here. Are we destined to simply repeat ourselves ad infinitum, with minor variations? For a generation of thinkers, artists and doers growing up as I write, intellectual boundaries lie around them in tatters. But unlike for previous generations, rebuilding those boundaries is just to repeat history. Subjectively, in terms of new big ideas, we've exhausted a seam, got stuck in a cycle; as with wonder, our intellectual surprise also hits diminishing returns.

There might be an equivalent in the harder sciences. The notion of convergence, of finding a unified story of everything, has been described as the 'deepest idea in the universe'. It offers a single, interlocking system of explanation, an all-encompassing synthesis taking us from the opening moments after the Big Bang to the present on Earth. It can tell us of the chemical properties of the universe (thank you, Humphry Davy and co.), the structure of distant galaxies, the movement of tectonic plates, the development of life including human beings. It can describe the growth of civilisation, cities, writing, take us through the many peoples and nations and cultures that have delivered us to the present, and explore our own minds and psychologies. Phenomena as diverse as language, the climate, biochemistry and industrial technology are all part of the same integrated system of explanation.

Given the success, it would be surprising if there were no diminishing returns here. The impact can be measured: as species are discovered, objects in space identified, portions of scientific discovery are subject in the words of one paper to 'exponential decay'.[53] It's the flip side of low-hanging fruit; less fruit on the tree, fewer new trees to find.

There are major gaps, of course: the origins of life, for example, the question of intelligent life, the origin of being itself. But the point is that compared to any time in history, knowledge has a remarkably complete superstructure; most scientists spend their time gilding an edifice that is already substantially built.

Knowledge is saturated. Markets are saturated. Cultural forms are saturated. This broad concept of subjective saturation shows how challenging it will be to keep coming up with new big ideas. Quite simply, the more you have in any given area, the more difficult further breakthroughs become – like wealth, there is a declining marginal utility to ideas. The first iPhone was a huge breakthrough, a new kind of device. A new smartphone is better, more widespread, but it's just another phone. The Google algorithm, like the iPhone, was a colossal breakthrough in consumer tech (and computer science); but today, with thousands of the world's finest minds working on it, PageRank improves at a relatively slower rate than its initial creation.

Even over much longer timescales this effect plays out. Arguably human wisdom or morality has not improved for thousands of years – it's hard to say we have decisively gone beyond Buddhism or the Stoics in this regard, and perhaps we never will or can.[54]

And there is a further mechanism: the more we have, the more we may overlook or misunderstand potential breakthroughs. Drowned in new things, we miss what matters; great science or art gets buried under the deluge of production. At some point the burden of knowledge becomes so great that a field like cancer research or the economics of innovation becomes too large for individuals or teams to comprehend. It fractures and siloes as a result.

Overproduction breeds problems. A paper on Alzheimer's research was published in *Science*; four years later, failing attempts at replication, it was retracted, but not before earning 500 citations. In the words of Dan Sarewitz, 'poor quality research metastasizes through the published scientific literature, and distinguishing knowledge that is reliable from knowledge that is unreliable or false or simply meaningless becomes impossible.'[55] This is what a saturated system looks like: incapable of self-correction, clogged, burdensome. No wonder there is a reproducibility crisis. Moreover, because there is so much literature out there, scope always exists for finding supporting evidence or ideas, however tendentious. As anyone writing a book of non-fiction can testify, the excess of academic production makes finding a paper to support whatever you might want to say easy. Not only do erroneous ideas percolate, but erroneous or misguided big ideas can warp entire ecosystems around them, blocking progress and new ideas from forming.[56]

Innovation theorists and economists suggest that new ideas and economic growth alike happen at the meeting point of disciplines and worldviews.[57] Technology and ideas are combinations, assemblages of other ideas. Given that, we should expect, over time, useful and interesting combinations to be tried out and discovered, again in a unidirectional process. A level of 'combinatorial exhaustion' is therefore likely: at some stage, whether now or in the distant future, making productive combinations will get more difficult.

Take painting. There are infinite ways you can apply paint to a canvas, in terms of colour, composition, form and subject matter. The subset of arrangements that are pleasing or interesting to humans is surely less than the total. It is still an endless space, and for as long as we exist it will be possible to produce new, beautiful or fascinating paintings. Yet the number of possible arrangements that will significantly alter art history or the way we think about representation and imagery, like the Impressionists or the Cubists, are much smaller still. A much more limited set of combinations of paint and canvas qualifies.

At some point the possibility that a painting will change thinking about the image is basically exhausted. I'm not suggesting we are there, or will be soon, but, as a thought experiment, I am suggesting that this is possible, and that it is more likely to happen as time goes on. Over the next century, with its supercharged production of visual imagery, we will, bit by bit, reach a point of combinatorial exhaustion in art historical terms. New contexts will bring new controversies as norms change and are transgressed. But in the fundamentals of painting, the chance for radical originality, for producing the unseen or unimagined, is less likely as visual saturation is approached.

It may be that there are infinite combinations, but there are not infinite productive combinations, and there will be even fewer truly significant or game-changing combinations. Absent some radical change, we move ever closer to a point of declining combinatorial possibility.

Throughout history the collision of cultures has been a fecund ground in this regard. Consider how the European Renaissance was sparked by trade with East, allowing the recovery of lost works, or how the Columbian Exchange following the connection of the Americas with Eurasia sparked a pan-continental transformation in everything from diet to economics. Yet globalisation has homogenised the world. We wear the same clothes (jeans, suits and baseball caps), eat the same foods (pizza, noodles, beer, cola), work for and buy from the same companies (Apple, Walmart, VW, Nestlé) watch the same films (Disney), speak the same languages (English), worship the same idols (TikTok, Pokémon, the Beatles). Societal interchange has become so immediate and total that everything blurs into a single culture where only the far interstices are free of the bland blend. Exchange of ideas and cultures will inevitably start to diminish. We will, at a global level, experience a kind of cultural combinatorial exhaustion. There may be reactions, but 'Sooner or later, on present showing, we shall have only one worldwide culture. So we shall have no one to exchange or interact with. We shall be alone in the universe – unless and until we find other cultures in other galaxies and resume productive exchange. The result will not be the end of ideas, but rather a return to normal rates of innovative thinking.'[58]

Normal rates in this context means the rates that dominated since ancient times, and by our reckoning those are very slow indeed. Again, this isn't just about low-hanging fruit being picked; it's a drawn-out, centuries-long process of exhaustion.

Just as a nation's economic growth slows down once you reach the frontier of technology, so – once you become habituated to large changes, make use of the most obvious combinations or approach the limits – with the ideas frontier more widely. Ideational progress is like travel: as one hits higher and higher speeds, resistance increases. Like a body approaching the speed of light, incremental gains become slower and more costly. Maturity is a drag.

As the frontier recedes it becomes more difficult to reach, tougher to crack. Gains in pushing it back cannot simply be repeated.

These processes will intensify over the next decades and centuries. In some areas we have started bumping up against limits. In others the complexity or depth of the problem has superseded our present capacity to grapple with it. Those problems will only get more complex, more intractable, as we approach the edge of possibility. Without the entry of radical new areas of enquiry, we inevitably get stuck. And even when new fields do open, it won't change the burden of knowledge or upend the psychological treadmill. This all adds up to the set of overlapping mechanisms I call the Idea Paradox. Big ideas in the future will only get harder.

To practitioners, saying that feels like a copout. It's like admitting defeat when those of yesteryear just raised their game. And surely, stepping into the unknown was never straightforward or comfortable. Yet ignoring the shape of intellectual history isn't going to help. The lesson of the Idea

Paradox isn't that everything is terrible or they had it so much easier in the seventeenth century; it's that we're going to need bigger fishing rods. Far from an admission of defeat or inadequate recognition of past challenges, the Idea Paradox clarifies that we need an emboldened, confident response.

6

The Stagnant Society

Inventing the Twenty-first Century

For forty-plus years, Bell Labs, the greatest industrial laboratory in history, 'the idea factory' of the twentieth century, invented the future. Then it stopped. Why?

If one big idea defines the early twenty-first century, it's digital technology: a world-changing general purpose technology. But the story of its creation is revealing. The Digital Age was spawned in Bell Labs, and in other very specific places like Bletchley Park, Princeton, the state-run Advanced Research Projects Agency (ARPA) and Xerox PARC. Without these organisations it's unclear how we would have reached our present technological settlement. The production and dissemination of big ideas can be fragile, contingent on specific circumstances. Change them just a bit and the breakthrough could be lost.

Thirteen of Bell Telephone Laboratories' researchers won Nobel prizes, most recently in 2018, for nine separate discoveries. From small beginnings in New York, Bell's inventions can scarcely be overstated. They include the transistor, a semiconductor which was the foundation of the entire digital universe; not to mention the solar cell, quartz timekeeping, radio astronomy, the laser, satellite communications, the mobile phone network, information theory, the UNIX operating system and programming languages like C and C++. Bell Labs was even pivotal in the discovery of cosmic microwave background radiation, the central evidence for the Big Bang. It's scarcely an exaggeration to say our communication and information infrastructure, the blueprint of the twenty-first century, is built on Bell science and technology.

Its West Street lab in Lower Manhattan and its spacious New Jersey campus were at the centre of a seismic change. At its height in the 1960s Bell Labs had over 15,000 employees, including 1200 PhDs and some of the most extraordinary scientists, mathematicians, engineers, technologists and managers of its time.[1] The lab was not only a scientific and engineering success, but laid the groundwork for a new commercial era, both for its parent, AT&T, and the world at large. Born out of AT&T's monopoly on American telephony – not only a vaunting commercial ambition to own the world's largest telecoms market but also a technological one – Bell Labs pursued a seemingly distant and near-impossible vision of universal connectivity.

A resulting tension ran through the labs: on the one hand Bell was part of one of the world's largest and most ruthless monopolies, famous for crushing rivals, and on the other it was a place where wilful eccentrics were given latitude to follow obscure lines of enquiry. For decades Bell Labs was in effect a shield for AT&T – its research was shared with other companies while the main monopoly was left untouched. While 'Ma Bell' was the epitome of a conservative, predictable behemoth, Bell Labs fizzed with unpredictable creativity. It was an odd combination, but in the fertile soil of the mid-twentieth century, it worked.

Over the years it employed an extraordinary cast of talent. Some of the more famous names include the combustible and abrasive Bill Shockley, Solid State Physics Group Leader in the late 1940s, part of the team that invented the transistor. He was later instrumental in founding Silicon Valley before descending into spurious race science. His colleagues, John Bardeen and Walter Brattain, had made the breakthrough itself: a relatively crude semiconductor device of gold foil, a germanium crystal and a paperclip, designed to boost or switch electronic signals. The three of them won the 1956 Nobel Prize in Physics.

There was the ethereal but genial Claude Shannon, progenitor of one of the twentieth century's most significant insights: information theory, not only an underwriter of the digital age but now posited as a fundamental property of existence. There was a cadre of visionary administrators, like the steady hand of Missourian Mervin Kelly: capable of balancing the competing needs of a demanding corporate parent and wayward scientists in a mushrooming organisation working at the limits of technology. In the words of a colleague, he was a 'supernatural force'.[2]

Kelly began to talk of the lab as something new, 'an institute of creative technology',[3] redefining his role not as an inventor but as the inventor of systems of innovation. Kelly, Shockley and Shannon were members of a group known as the Young Turks; others were innovators like John R. Pierce, who worked on the first communications satellite *Telstar 1*, and James Fisk, later a president of the lab. The Young Turks were given free rein to take risks. Here was a scientific community whose excellence and freedom rivalled or exceeded the great research universities while remaining close to the potential for practical change. Even Bell Labs executives privately marvelled at their freedom. Only about 20 per cent of the labs' staff worked on basic research, but this still meant thousands of people were engaged in what one contemporary analyst described as 'idle curiosity' – not at a university or for private study, but in a profit-driven corporation.[4]

Eventually that tension was to prove too much. Over the years came regular calls for the breakup of the mighty Bell 'system'. It clung on for decades, but even the labs' phenomenal output couldn't save it in the end. By the early 1980s the Reagan administration was no longer prepared to tolerate such a monopoly. A legal tussle saw an agreement between AT&T and the Justice Department: the mothership would be divested of its local phone companies. On the face of it this didn't impact Bell Labs, but the die was cast for a new kind of organisation, one capable of supporting a conventional industrial laboratory – just not a world-changing 'idea factory'. Today it remains a leading example; now owned by Nokia, it continues to win accolades for its work since the 1980s. But no one expects the next fifty years to be defined by Bell like the last fifty.

In some ways this wasn't bad for innovation: market competition saw the rollout of new cable, which in turn enabled and boosted the Internet revolution. It was less good for making fundamental leaps like the transistor or information theory. With a handful of exceptions, the journey is typical of industrial research more widely, symptomatic of how a change in capitalism meant that 'idle curiosity' was pushed out of the marketplace.

The story of Bell Labs suggests that our bases for new ideas are narrowing, growing more pressured and insecure. The industrial research which powered so much twentieth-century invention is, outside a small group of tech companies, a shadow of its former self. Bell was always part of a much wider ecosystem of corporate innovation: started by Edison and German conglomerates, by the 1910s industrial labs owned by the likes of DuPont and Kodak were pioneering radical new products. DuPont's Experimental Station went on to develop nylon, neoprene and kevlar (among others), ushering in the era of synthetics; in the 1960s the company published more in the *Journals of the American Chemical Society* than MIT and Caltech combined.[5] Later labs run by the likes of Merck, Dow, Lockheed Martin (the legendary Skunkworks) and Xerox did the same in their fields. And it wasn't just Bell who won Nobel prizes: so too did researchers at companies like General Electric, IBM, Texas Instruments and Sony. IBM's Fellowship programme, established by Thomas Watson Jr in the 1960s to produce what he called 'wild ducks', untameable minds that in the words of a later advert were 'dreamers, heretics, mavericks, gadflies and geniuses', has produced the Fortran programming language and the first devices to image atoms. It can boast five Turing Award and five Nobel winners among a string of accolades.

This broad-based ecosystem has eroded. One senior Bell Labs researcher, Andrew Odlyzko, making sense of what was happening to his institution over the 1980s and 1990s, called it 'the decline of unfettered research'.[6] The image is significant – society was placing institutional chains around the curiosity-driven research agenda that enabled people like Claude Shannon to explore information theory. Research had been commodified into narrow, directed chunks. Bell had changed; 'rational' economic decision making was in. Writing in 1995, Odlyzko had already seen the writing was on the wall for the industrial labs that had nurtured his career: 'the prospects for a return to unfettered research in the near future are slim. The trend is towards concentration on narrow market segments.'[7]

Research from MIT shows that while companies still invest

in innovation, the focus is on practical applications rather than basic science.[8] In essence they piled more into R&D, but forgot the R in favour of the D.[9] Only around 7 per cent of corporate R&D is spent on basic research and 16 per cent on applied research – the rest is all development.[10] While company R&D still accounts for a large portion of total global R&D spend, it comes in at under 20 per cent of the funding for basic research, even as government funding for such research as fallen in relative terms in countries like the US.[11] What used to be a team effort – public and private working together – has been dumped on government.[12] Business indices, from R&D awards to publications in top journals, all show marked downward trends; the number of AT&T's publications alone dropped 81 per cent from the 1990s to the 2000s as its research budget was decimated and corporate emphasis on publication deprioritised.[13] Indeed, since 1980 there have been marked falls in corporate publications across the board, especially in 'basic' or 'influential' areas.[14]

For big business, developing existing ideas in a safe, predictable fashion has become the priority.[15] The notion that companies should engage in 'idle curiosity', that this might be a significant aspect of their activity and future success, has grown anachronistic.

Business trumpets its capacity for innovation even as it channels resources into the most cautious forms of innovation. The economist Mariana Mazzucato argues that cheerleaders for innovation like Google, Apple and big pharma are in fact reliant on government-built technologies. The basic technologies behind the iPhone, like GPS, capacitive touchscreens, voice-enabled assistants and Internet connectivity all relied on government-funded support; Google's core PageRank algorithm even acknowledges it on the patent. Developing new treatments for disease requires huge grants from the National

Institutes of Health or similar; only once foundational work is established do startups and corporates step in. The businesses of today don't invent the technologies of the future; instead (not that this is to be sniffed at) they integrate them into desirable and easily consumable products.[16]

A shift in capitalism has changed corporate imperatives. In an age when maximising shareholder value is the priority, a new breed of company has less time to invest in wider social goods, which explains much of this decline. Spillovers from basic R&D benefit society more than they do the originating company: William Nordhaus estimates that only 2.2 per cent of the surplus from money spent on knowledge accrues to the investing company. Another study suggests that 57.7 per cent of the gain accrues to society rather than as rent to the company (which sees only 13.6 per cent of the gain).[17] It was famously not Xerox that profited most from its legendary PARC lab, but Steve Jobs and Apple, who saw its graphical user interface on a tour and went on to launch it as the benchmark in useable software. Even in the very best of times knowledge is underproduced and invested in thanks to this free rider problem.

Moreover, in the heyday of Bell Labs corporate profits and executive pay were highly taxed; you might as well spend on R&D. Today the reverse is true. Why put money into R&D when it can be safely funnelled back to shareholders or the C-suite bonus pool?

This all has real consequences: the decline of the corporate research lab has meant science is removed from the economic front line. Even if you take the more positive spin on all of the above – that corporates have simply outsourced research to universities and startups – materially fewer innovative products and processes are created.[18]

Society itself is stuck, working against radical ideas. Despite the boosterish pronouncements of Silicon Valley gurus, we are hostile to breakthroughs, more conservative, cautious, rule-bound, anti-maverick, return-oriented, ideologically blinkered and short-sighted than we might think. There are no easy answers to the question of how this could change, which is why the social context weighs so heavily on the future, compounding the Idea Paradox.

Bell isn't unique. Rather it's emblematic of a shrinking of corporate horizons back to the bottom line: a warning of societal changes that are stifling our capacity for big thinking.

History doesn't work on a single straight line. It's quite possible to make progress on some fronts while seeing reversals or obstacles in others. It's equally possible that, even long past an apex, it feels as if the world is rushing forward, the golden age only becoming apparent once the glow fades. One recurrent big idea is the sense that history is cyclical: that times of plenty and promise are followed by downward turns. A generation's problems are laid down in the solutions to older problems long since forgotten.[19]

The German thinker Oswald Spengler agreed. An extreme pessimist writing in the ashes of the First World War, he decried any idea of unidirectional progress, seeing in his own era a parallel to the fall of Rome.[20] Spengler argued that societies and cultures have, like organisms, a natural life cycle. First early, stumbling periods; then a phase of creativity and maturity; and then, inevitably, a period of decline and crisis: in his words, a winter. For the overconfident West, Spengler was convinced that the good days were over. Eventually, slowly, civilisations hit a 'decadent' phase once the decay sets in. He believed this twilight was here, and would only intensify.

Spengler is one of those lugubrious sceptics whose frankly absurd historical categorisations are largely forgotten today.

But his schema of societies moving from periods of creativity to stagnation has force. He might be a crank, wrong about the timelines, but nevertheless on to something with the bigger picture. Based in Munich, he witnessed one such transition almost on his doorstep, one of the greatest stories of intellectual boom in history: Vienna.

In the words of *The Economist*, the twentieth century was 'the Viennese century'.[21] Just as Bell Labs built the technology, so Vienna built its mental architecture, an idea factory on the scale of a city. It may have been the dog days of the Austro-Hungarian Empire, but *fin de siècle* Café Central or Café Landtmann thrummed with mental and artistic adventure.

Extreme creativity was everywhere in late Habsburg Vienna. This was the world of nascent psychology and the unconscious, studies of dark, primeval urges and unleashed sexuality; the clean lines of modernist architecture met the twisted canvases of a new style of painting. It was the birthplace of modernism and the bracing philosophy of logical positivism, of Germanic fascism and neoliberal economics. Vienna created, for better and for worse, the modern world.

It was here that Freud invented psychoanalysis; Schoenberg atonal music; Adolf Loos modernist architecture. Ernst Mach explicated the physics of shock waves; Theodor Herzl established Zionism. Vienna was the city of Gustavs Mahler and Klimt; a medical and a philosophical centre; the milieu of Robert Musil, Stefan Zweig, Arthur Schnitzler, Alfred Adler, Egon Schiele, Ludwig Boltsmann, the controversial painter Oskar Kokoschka, the young Ludwig Wittgenstein. It was also astonishingly, darkly, the crossroads for Trotsky, Lenin, Tito and Hitler – a stark reminder that big ideas are not always to be welcomed. Military defeat in the First World War may have brought down the empire, but it unleashed a last wave of energy. Charlotte and Karl Buhler made a systematic study of child development. The Vienna Circle, with its brilliant leading names like Rudolf Carnap and Otto Neurath, developed one of the most distinctive and influential philosophical movements of the twentieth century.

Out of this too came Karl Popper, perhaps the century's most significant philosopher of science and a major proponent of political liberalism. This émigré generation were nothing if not big thinkers: Peter Drucker originated management theory; Paul Lazarsfeld established sociology in America; economists of the Austrian School like Friedrich Hayek, Ludwig von Mises and Joseph Schumpeter set the neoliberal foundations for the modern discipline and thus, arguably, for society and capitalism.

Throughout these years Vienna was a city of contradictions, riven with anxieties that gave life to this ferment: at once ordered, bourgeois, hard-headed and rational, politically illiberal, monarchist and stolid, and at the same time a seething mass of political and aesthetic transgression.[22] The world described by Stefan Zweig is one of Hungarians and Czechs, Jews and Catholics mingling in coffeehouses and bouncing around challenging ideas. It was an imperial capital, but with eleven distinct peoples and nationalisms from Ruthenes to Romanians, Slovaks to Slovenes. In 1910 its two million people made Vienna the sixth biggest city on earth and it industrialised fast through the late nineteenth century, throwing up fortunes like that of the Wittgenstein family, even as it remained mired in traditionalism. It was big enough for critical mass but small enough that ideas were quickly shared. It was hierarchical and traditionalist but old boundaries between topics were ignored by an avant garde, and new areas, like sex, made the subject of enquiry. It was a culture of, at once, science, books and learning, Babel-like polyphony, individuality, a cacophony of identities; a place to break down old categories even as an old guard and an absolutist monarch tried hard to impose them.

Vienna's specific circumstances weren't to last. By the time Freud left in 1938, hounded out by the Nazis, the energy was spent. Just as with Bell Labs, no one suggests that the twenty-first century will be another 'Viennese century'. Here is the kicker: Vienna today is, in practically every sense you care to measure, a much improved city compared to the era of its intellectual boom. Visit Vienna and you will find it endowed with *Kultur* and *Kuchen* alike, wealthy, beautiful, successful, a place which regularly tops world quality-of-life surveys. It has strong universities, research institutes, international bodies and businesses. But it is not one of the boundary-pushing metropolises of the twenty-first century. If anything it marinates in nostalgia and small-c conservatism. If you had to work or live in the Vienna of 1905 or 2025, you'd go for the latter. But if you wanted to have an idea that would change the world? Easy.

We don't need to go full Spengler to wonder whether there is something of Vienna about the entire Western world, a Viennaification of our ideational environment; not so much the fall of Rome as a mix of self-satisfied embourgeoisement, timidity, complacency and pointless rancour – even in the face of, say, a global pandemic. Like Bell Labs, blessed with a glorious legacy, a great foundation and much forward momentum; like Bell Labs, like Vienna, slouching past its most revolutionary phase into something altogether more 'decadent', more comfortable, more indulgent.

It's still probably better to do almost anything today than in the great age of Bell Labs or the Viennese heyday, let alone the seventeenth century. Although the ground may look fertile, however, all kinds of weeds and poisons accumulate nonetheless. As a result, we're often content to dodge the challenge of the Idea Paradox; to invest in D and not R; to erect structures and processes that impede ambitious thought.

How we approach the next generation of breakthroughs will turn on the three forces that define a stagnant society. First, we look at those of excessive financialisation.

Mammon's Empire

Around 1980 capitalism changed. Building on the work of Austrian School economists (of course), a series of reforms coupled with changes in corporate thinking made markets more open and deregulated, their mechanisms absolute. A new model of shareholder value gained precedence. So far, so familiar. What is less often discussed are the consequences of this shift for our ability to think anew. If having an idea was simply a matter of sitting in a room, none of this would matter. But it's not. Bringing an idea into the world, whether launching a new cancer treatment or publishing a book, requires resources. And resources is another word for money.

It was assumed that unshackling markets and entrepreneurs would lead to an orgy of creative destruction. In fact, as we saw in Chapter 3, the results have been tepid. As the economy was subsumed into a kind of quasi-financial sector, new barriers arose: sunk costs and warped incentives project far into the future. As society became addicted to easy returns, quick-fix attitudes and beliefs ate into social and institutional structures. Business, science, cultural practice and government redrew the realms of the possible inside the conservative compass of prudent accounting and shareholder return.

Short-termism is rampant. Companies focus on the next quarterly earnings call; CEO tenure and stock ownership intervals are both on long downward trends, the latter from years in the 1940s to seconds in the 2010s.[23] A survey of over

1000 senior executives reveals they all feel acutely pressured by short-term factors detrimental to long-term success.[24] The development of big ideas requires time, space and money. In this world of fast-moving stocks and CEOs they get squeezed. A relentless focus on near-term monetary reward comes at the expense of costly and uncertain new ideas. In the early 1990s Boeing changed its corporate mission from building aeroplanes to delivering shareholder value (which, it should be added, promptly tanked). Should we be surprised at the subsequent lack of breakthroughs in aviation?

Earlier we saw how the West now has slower long-term productivity growth, lower and less successful business creation, more concentration among big established firms and lower overall growth. One key change, exemplified by Boeing's nineties shift, was in ownership, which was increasingly dominated by what two researchers call 'gray capital': pension funds, sovereign wealth funds, hedge funds ... Eighty per cent of US stock is held by such institutions, more in the UK. Ownership of the firms driving large-scale innovation is hence diffuse, devolved to a cadre of professional asset managers removed from immediate responsibility, lacking a clear vision or leadership and relentlessly focused on quarterly earnings. Again, the rational course for these managers isn't long, risky, costly new ideas but steady returns.

None of this leads to more dynamism. Firms invest much less, including on R&D, when they have institutional ownership and when a sector sees a great concentration of firms. When there are fewer but bigger companies, they are incentivised to return the money and don't worry as much about competition. No wonder we have a sclerotic and, ironically, closed economy. No wonder company net investment has declined. Corporate investment, much higher than noncorporate business investment, fell from 30–40 per cent in the period

1950–2000 to 20–25 per cent after 2000, and that ignores the dip after the financial crisis, to say nothing of the coronavirus recession.[25] Cash is not spent on risky ventures but hoarded or returned to shareholders in record amounts. Measured as a proportion of internal cash flow, returns to shareholders are six times higher than the 1970s.[26] S&P 500 companies alone have trillions in cash, splurging a record $800 billion on share buybacks, almost $500 billion on dividends, while Apple and Alphabet alone sit on hundreds of billions.[27] All this epic hoarding of cash is described as 'precautionary'.[28]

If capitalism used to be about owner-managers taking risks to create new breakthrough businesses, the interwoven, highly intermediated 'gray' structure of modern equity markets is the opposite. It's not that business doesn't have the money to radically change the world; it's that managers and institutional owners neither need nor are incentivised to do so. In other words, modern business practice stifles the very creative destruction that was meant to be the productive essence of capitalism.

Demographic effects play another role – research argues that the presence of more older people in the population translates into more concentrated and older companies, which in turns drags down dynamism.[29] As the population ages, pension funds accumulate more and more of the economy; they don't want disruption, they want reliable income. The older we get, the less risk the economy can bear. In general, one would expect an older society to be less dynamic, more risk- and loss-averse. Others see the role of competition, market barriers or access to capital as having the same effect. The capital required to unseat incumbents, whose market concentration has grown even as the number of startups falls, has risen. Ufuk Akcigit and Sina T. Ates see the role of those ever-stronger market-leading companies as critical.[30] They note increases in the companies' power and wealth, and corresponding increases in aggressive patenting activity (more of which below).

This bloating and ossification enables establishment firms to dominate markets without real change. Vested interests, always averse to the unknown, then cream off the rewards of their entrenchment. Rent-seeking trumps innovation. Many companies cosy up to government for easy contracts – crony capitalism is still here. Markets, including the commanding heights of tech, are gridlocked, the same blue-chip names paraded year after year. Incentives for wrenching change wither in this environment – indeed, sectors that have a near miss on a major merger or acquisition remain more innovative, see more new companies and investment come in.[31]

The results are the pronounced shifts, already observed, towards development and away from research; pressure to return earnings to shareholders rather than reinvest them; a consolidation of big, older companies unwilling to disrupt themselves; a requirement to prove financial return that infects everything from the arts to academia.[32] Over 80 per cent of the value of the S&P 500 is in intangible assets, which only makes companies want to husband and protect such assets, not radically alter them.[33] The holding of such assets, including software, has moreover been correlated with decreased dynamism: sectors associated with these investments see a greater persistence of firms and less leapfrogging.[34]

All this creates what Mazzucato calls a 'parasitic innovation ecosystem', where large, powerful incumbents coast on work done elsewhere.[35] Away from the glossy brochures, the phalanx of buzzword-spouting management consultants, the calendar-filling conferences, is a system both short-termist and static, risk-averse and returns-driven, timid and defensive. Moreover this hyper-financialism seeps into government policy, cultural centres and independent research institutes, all of which are subjected to its methods and imperatives. It sets the institutional weather.

Venture capital (VC) is no panacea. In fact it's often part of that same parasitic ecosystem.[36] Timescales in biotech or materials science are often too long for venture capital funds to realise a return. Creating a minimum viable product is, unlike with software, capital and infrastructure intensive. Cleantech is a good example. In 2006 $1.75 billion was invested in cleantech. By 2011 that amount had boomed to $25 billion.[37] But then at least half the money was lost in company failures. Venture capitalists shied away from such investments as they showed that 'real world' tech was more difficult and longer than the cycles of software and traditional VC investment. Cleantech was damaged, progress set back years. VC-backed R&D struggles with this kind of physical prototyping, which, unlike websites, comes with enormous startup costs and steep learning curves. Venture capitalists often step in quietly once the most risky foundational work is complete. Even then the support amounts to a fraction of the whole – only around 0.2 per cent of startups get VC backing, and VC funding is still tiny compared with demand, total investment or the economy as a whole.[38]

This hyper-financialised world has other insidious impacts. Take intellectual property (IP). Granting time-limited monopolies in the form of patents or copyright, IP should stimulate invention and creativity, enabling creators to benefit from risk taking. Again, in the 1980s and especially in the US, the system changed.[39] It became easier to get and enforce patents. The United States Patent and Trademark Office adopted a model where applicants paid, at a stroke incentivising grants as a revenue stream. But it was also starved of resources, losing the expertise needed to adequately assess applications. Patent offices cannot compete with either the salaries or the expertise of large firms, particularly in complex and fast-moving sectors. Even as they were pressured to make money, patent examiners were overwhelmed by a flood of applications.

The result? There was an explosion in the number of patents granted, tripling annually between 1983 and 2002.[40] Patent trolling and squatting were normalised. Companies engaged in internecine patent arms races. Small companies, particularly in areas like software or biotech, couldn't compete and were locked out of the market.[41] Patents became offensive weapons, hoarded as stockpiles to attack competitors and extract rents. Fear of litigation – eye-wateringly expensive in IP disputes – was enough to block new endeavours.

Patents also became increasingly absurd: Amazon famously patented a 'one click' shopping method. A five-year-old was able to patent a 'Method of Swinging on a Swing'.[42] Polaroid was, for years, able to put anyone creating an instant photography company out of business.[43] A jam company patented what were, in effect, toasted sandwiches, and indeed the act of toasting bread.[44] British Telecom claimed to have patented the hyperlink, which in theory would allow it to claim royalties from every user of the Internet.[45] There will always be a balance between encouraging and inhibiting new ideas within the IP system, but IP flipped from a cornerstone of that system to an impediment.

As the world of 'gray capital' and financialism colonised the economy, it skewed rewards in its favour for talented individuals. Developing a cure for dementia or exploring new energy sources is a less tempting prospect when you can earn millions moving other people's money around as a hedge fund manager. It's more lucrative and straightforward for engineers to build derivatives-trading algorithms than to design a generation of carbon-neutral airplanes. One study suggests that, of a cohort of Stanford MBA students, those going into finance earned three times as much as those who didn't.[46] And that was Stanford MBAs, hardly low earners. For skilled people it's never looked so comparatively unattractive to pursue a career as a physicist or theatre producer. A perverse fact of the modern economy is that most of the world's best and brightest spend their productive years staring at screens, writing emails and compiling spreadsheets on work that is at best historically irrelevant. Our economy rewards trivia and ephemera over originality and long-term impact.

It goes beyond finance. If you care about big ideas, it's clear that the world misallocates talent on a mind-boggling scale, whether our best minds are peddling ad space for Facebook or M&A advice at Goldman Sachs. The most significant brain drain in the world isn't between countries but away from the frontiers. Population, education, wealth, technical capacity are all indeed up – but they are also directed towards activities with little lasting value.

If you care about big ideas, a market militant ideology is not ideal. That's not a point about big business as such. After all, both AT&T and Google are big and innovative. Nor is it a problem with markets, which have been powerful and essential drivers of ideas throughout history. Rather it is about the kinds of businesses and markets – and, more importantly, the innovation ecosystems – we have created, sustained and project into the future.

Whether it's making a computer game, writing a novel or building a flying car, society's structure of incentives doesn't favour the slow, difficult, uncertain, unruly conditions entailed in doing something radically new; it favours the rentier and the safe. And that's without taking into account the impact of some further financial calamity. When or if it does hit, such a crisis will deprioritise exactly the expensive, risky and difficult work that is already facing hurdles. And, unfortunately, financial blockers are not alone.

Please Tick All the Boxes

The brisk new capitalism was meant to blow away stuffy post-war bureaucracies. That didn't happen. Instead whole new layers of bureaucracy accumulated and continue to do so. We live not in a freewheeling and experimental age, but rather in a 'utopia of rules' subject to an all-pervasive managerialism.

If anywhere was a bulwark against the necessity for speedy returns, surely it was the flourishing, much-vaunted university sector: the home of foundational research and eccentric ideas alike. After all, universities have never had it so good; they're not just storied research centres but supposedly core engines of our cultural, technological and economic activity.

Yet everyone I spoke to in academia, from feted professors to newly minted postdocs, felt something close to despair at the system. To them the notion that the ivory tower still exists as some last redoubt against the petty, frothy world outside seems comical.

Over recent decades, the character of universities has changed. Even as elite institutions have become richer and more powerful, so they have aped the blue-chip corporations dominating the economy – with all their attendant vices. Driven by a cadre of well-remunerated professional administrators, they have lost sight of untethered ideas. Between 1975 and the twenty-first century those managers and administrators have proliferated. At public universities in the US their numbers are up 66 per cent. At private universities it is, incredibly, more than double that rate: a 135 per cent increase. Both figures are hugely in excess of faculty and student growth, leading to the creation of an 'all-administrative university'.[47]

Indeed, administrators and 'other professionals' outnumber actual faculty members.[48] Since the 1970s spending at universities has tripled, but given the faculty–student ratio has stayed the same it's questionable where this money goes. In my conversations this was regarded as at best wasteful, at worst actively fostering a grim box-ticking atmosphere inimical to big thinking.

No one outside the offices of presidents, vice-chancellors, provosts, senior vice presidents, deans, associates and assistants is quite sure how this supports the world of ideas. Instead spending on student services over a decade is up more than double against spending on teaching and research.[49] Investment is channelled into sports stadia, gyms and luxury leisure facilities rather than towards labs, experiments and researchers. Departments of philosophy or languages or the arts close while new dining halls and shops spring open.

Senior managers care about rankings and public profile, as audited by the media, government or internally. In practice, every elite university spends enormous energy securing a good place in the global rankings. In the UK every university and academic must justify their existence with extensive exercises that grade every layer of research and teaching. Universities have thus become classic victims of Goodhart's Law: that when a measure becomes a target, it ceases to be a good measure. This doesn't stop them adding more such statistical measures all the time, and nor does evidence that an over-reliance on them dampens creativity.[50] And so the corporate world's audit culture has ballooned into the dominant fact of life for universities, where researchers spend a vanishing portion of their time on their basic job: instead they write proposals for grant money, assess those proposals, sit on committees, write reports, fill in forms. Much activity seems to involve marketing and PR for the course, department or university.

Just as in corporate life, endless guff about vision and innovation replaces actual practice: in the words of one critic, 'In the all-administrative university it is entirely appropriate that mastery of managerial psychobabble should pass for academic vision.'[51]

The independent scholar Alexey Guzey highlights another example.[52] When the Harvard Fellows were created in 1933 they were 'all chosen on the basis of their exceptional promise, not their past work' to 'provide an alternative [to the PhD] path more suited to the encouragement of the rare and independent genius'. Harvard wanted oddballs who could operate outside the usual channels. But by 2018, thirty-eight out of thirty-nine Junior Fellows were either PhDs or enrolled on PhD courses. So much for encouraging 'independent genius'. Instead only those same-old channels passed muster. It is symptomatic of credentialism (a much wider affliction), and of how non-traditional approaches are discounted. If even the Harvard Junior Fellows all fit the mould, what hope elsewhere? Scientists win acclamation on narrow metrics: their postdoc positions, their h-index. Even authors or visual artists are now expected to acquire a litany of postgraduate degrees.

Research gets caught in a vicious cycle: funders and universities feel that risky research or researchers just aren't worth it. In the developed world the academic system has hit maturity and stopped meaningful expansion, despite churning out PhDs. As a consequence jobs are scarce, competition fierce and zero sum; one wrong move can finish a career. Researchers, and particularly early-career researchers, feel obliged to conform and play safe. Every scientist I spoke to mentioned this unprompted: 'dangerous' blue-skies research simply doesn't get funded. Instead, where it happens, it's smuggled in through the back door.

A troika of box-ticking exercises dominate the life of the

mind: promotion and tenure; research grant applications; and peer review for prestigious journals. These might appear to represent essential quality control. Except all can punish the hubristic crime of thinking big, breaking norms, challenging establishment wisdom. These metrics govern the lives of academics, and yet each places their output in the hands of a risk-averse system. Everyone recognises the flaws of grant applications or peer review at major journals, yet there's no consensus on what to do about it. And so the status quo grinds inexorably on.[53]

You might think this is classic disgruntlement in the face of a changing world. But it is having a marked impact on what ideas people chase, and how they chase them. Analysis of large bodies of research shows 'a growing focus on established research at the expense of new opportunities'.[54] Conservative research strategies, burrowing deep into and consolidating existing knowledge, dominate the production of knowledge. And yet the research also demonstrates that more adventurous work leads to more discoveries and prizes. So why not go for it? There's a simple calculation: 'Research following a risky strategy is more likely to be ignored but also more likely to achieve high impact and recognition. While the outcome of a risky strategy has a higher expected reward than the outcome of a conservative strategy, the additional reward is insufficient to compensate for the additional risk.'[55]

The culture of metrics and citations is very much part of this shift towards safe, predictable work. In an important paper on scientific stagnation, Jay Bhattacharya and Mikko Packalen argue that the obsession with citations explains a huge amount of scientific stagnation.[56] Recall that research careers are dependent on garnering large numbers of citations. This creates a powerful incentive to chase them. But what gets citations? You might think it's the risky high-impact exploratory stuff. No – in fact 'incremental me-too science' tends to do much better, even as the publication system exhibits an overall bias against extreme novelty.[57] As more people work on a topic, so publications are cited more often, so more people pile in, and so on.

Exploratory work meanwhile often fails and gets overlooked – and even where it does eventually pay off, this can happen decades later, too late for the originators to benefit. For example, polymerase chain reactions are now an essential part of any science involving the analysis of DNA, including the widely used tests for Covid-19. It relied, however, on exploratory but seemingly obscure research looking at how bacteria survived in hot springs at Yellowstone National Park, work considered of marginal interest and little cited when published. But without it, the human genome project and the edifice of contemporary genetics would look very different.

Such stories are legion in the history of science. Surveying the field today, Bhattacharya and Packalen find a pronounced clustering of effort away from exploratory and breakthrough work into the incremental portion of research. Hence stagnation. What's more, the emphasis on citation and publication distorts research by incentivising behaviour that twists results and studies to favour whatever is likely to achieve both.[58] Disinterestedness gets lost.

Even funders acknowledge that the research ecosystem has become more short-term and conservative, siloed and fragmented, less adventurous and idealistic.[59] It gets trapped in three- to four-year cycles; after a point researchers must turn towards whatever will secure their next batch of funding. Indeed, up to 40 per cent of faculty time is spent chasing grants.[60] Evidence also suggests that funding evaluators are systematically biased against research that is either novel or close to their area of expertise.[61] The net result is that between the machinery of funding and the tribalism of peer review, the 1 per cent of research that could be a big idea is deprioritised, quashed by cultures of excessive accountability, administration, gaming and even fear. In its place is a drift to the middle.

Eccentrics and outliers are ever less welcome. As physicist Lee Smolin puts it: 'It is a cliché to ask whether a young Einstein would now be hired by a university. The answer is obviously no.'[62] What about a young Marie Curie? Or more recent mavericks like Stuart Kauffman, Lynn Margulis or David Deutsch (the pioneer of quantum computing who recently admitted his work would never have been funded today)?[63] Would a thinker like Derek Parfit, who published two of the most important books of ethical philosophy of modern times and little else, find a foothold in a system that relentlessly prioritises quantity? The truth is, universities are anti-maverick. If you want to challenge norms, the academy, much like the wider world, is no longer welcoming.

For administrators this works just fine; originality sounds better on a press release than in the lab or lecture theatre. Remaining oases like tenure, which guarantees academic autonomy, are whittled down: in the US only about 30 per cent of faculty is tenured or on the tenure track, compared to 67 per cent in the 1970s.[64] Humanities courses are down from one in five degrees in the 1960s to one in twenty, with the result that departments and liberal arts colleges are closing in droves – numbers of the latter have decreased by 40 per cent since 1990.[65] Swathes of academia are felled by the year, spaces of thought, education and expression gone. Meanwhile a Wellcome Trust investigation of research culture found that 75 per cent of participants felt creativity was stifled and 23 per cent felt pressured to deliver a particular result.[66] Only

29 per cent felt secure pursuing their career, and just 14 per cent believed the emphasis on metrics helped research. It paints a world dominated by patronage, pressure, careerism and frustration, with little scope for autonomy or originality. In that world freedom of thought is valued on paper – but in reality departmental politics, promotional opportunities, over-specialisation, funding cuts, citation-chasing, intra-disciplinary reputation, epistemic territory grabbing and the pressure to publish are omnipresent. Cultures of bureaucracy and narrow horizons rule in practice.

In the words of one paper surveying the whole field, this ecosystem is 'disastrous' for breakthrough ideas.[67] The managerialist culture is, concretely, moving us away from discovery and originality. As with the pressures of financialism, it sends ideas in one direction: towards the risk averse and small scale. And this is to say nothing of the power of orthodoxy or the seductions of fashion and groupthink.

Of course, managerialism isn't confined to universities. Corporate and government life is no better. As anyone who has worked in a large company can testify, managing, monitoring, discussing and reshuffling organisational complexity itself becomes a prime goal. Corporations are far from the lean innovation machines of myth: emails, internal planning, politics, arse-covering and favour-currying all take precedence. Dissent is discouraged, despite all the business books that say otherwise. And all the while management wastes time: an estimated 60–80 per cent of it on writing reports and in pointless meetings.[68] The Boston Consulting Group 'index of complicatedness', a measure of firms' internal complexity, has grown 7 per cent annually for an extraordinary five decades.[69] One survey of executives working on innovation found that their biggest barriers were not resources, but internal politics and culture.[70] Once again, the appetite for original thinking is diminished. For what it's worth, this applies not only in more staid sectors, but also in the great technology or entertainment businesses.

Public bodies are naturally stuck with the same managerialist pathologies, often in extreme form.[71] Governments profess profound support for new ideas, but tend to baulk in practice, so that innovation gets stymied at the execution stage. Meanwhile regulations, licences and standards accumulate. This metastasising thicket of rules and regulations mean new ideas must be tortuously squeezed and overlaid on top of them. The space for possibility is slowly suffocated by an acephalous process where special interest groups shout loudly, get their way and slather on legislation to protect themselves. Repeated again and again, the process crushes space for manoeuvre and experiment.

Go back to transport (more on which shortly). We could be on the cusp of a revolution but the necessity to navigate reams of regulation and changing policy may make change impossible. Autonomous cars, flying cars, tunnels under our cities: all are doable but regulators are (understandably perhaps) nervous. We saw the first signs of this with Uber: Paris has only 18,000 official taxis and 19,000 minicabs, against London's 23,000 black cabs and 40,000 minicabs ... And yet Uber has been fiercely resisted by the city authorities. What hope for flying cars? Or energy – recall that available energy growth has slowed. It's little wonder when regulations relating to new energies (like nuclear) are so fundamentally inhibitory.

After years of decrease up to the early 2000s, we are in a period of increasing regulatory burden: one in four American jobs requires some kind of official occupational licence. It's no secret that the number of lawyers has been growing fast for decades, increasing nearly fourfold in the US since the early 1980s. Between 1997 and 2012 the US Code of Federal Regulations grew by 12,000 restrictions every year.[72] It now runs to over 175,000 pages, much of it self-referential and opaque.[73] Years after they were declared safe by its own studies, the EU still outlaws techniques like basic genetic modification. The General Data Protection Regulation (GDPR) was meant to protect privacy, but like many ill-thought-through regulations makes life for startups harder – and as a result helps large existing companies. The world's fastest growing jobs are roles like private security guard or bank compliance officer, hardly productive. In everything from tech to logistics, construction to banking, compliance costs now form a significant barrier to entry and favour already dominant interests. So much for the mythologised notion of 'permissionless innovation'.

Over-zealous, over-complex regulation favours incumbents, inhibits innovation and creates hidden costs in the form of paths not taken. Moderate, judicious use of the precautionary principle is needed, but 'judicious' is a tricky balance. I'm not against regulation per se, but from the perspective of analysing why big ideas are not introduced and scaled up as frequently as we might expect, it's undoubtedly part of the mix. Nor am I against management; after all, Mervin Kelly at Bell Labs was a manager and it was Bell's management system that proved so effective. But like regulation, an excess or misdirected effort of management is the problem.

All of this presses on the future, incentivising the maintenance of the status quo. Cultural and academic fields, no less than business or politics, have their vested interests. Even more than the Catholic Church, the impetus against Galileo began with Aristotelian rivals at Italian universities. Between them, financialism and managerialism make people defensive, protective of enormous sunk costs. Businesses are invested in their current products, brands and supply chains. Individuals have invested their human capital in lengthening education and fiercely competitive careers. Building authority and claiming territory takes decades. No one enjoys seeing it come crashing down. Think of an oil company or a senior doctor. As time goes on, they have only a greater sunk cost and more reason to protect its value. For incumbents, incentives to come up with new ideas don't increase with time. Instead the reverse is true; they are incentivised to maintain what they know and what buttresses their status. The more any organisation or person has accrued sunk costs, the more they will want to preserve what they have.

Both the financial and the managerial ethos hate uncertainty. Yet the generation, execution and purchase of big new ideas is intensely uncertain. For big business, education or government, there is little reason to back a deeply uncertain proposal or person over the scores of well-meaning and attractive ones. Better indeed for the manager, board, institutional investor or funding body to take the path of least resistance. Regulators don't get fired for being too cautious and CEOs don't get fired for solid growth; real disasters hit home more than vague avenues of missed potential.

We have built a world where the fastest growing jobs are not in research or the arts but in regulatory functions: compliance not science. A world where inventions and medicines are never trialled, where the default is 'no' and risk aversion an art form, where comfortable incumbents can sail on unimpeded, where a 'tyranny of metrics', a thicket of rules and an over-laboured management structure dampens the scope for the new. These forces have built up over decades, seemingly impossible to dislodge or resist.

'Thinking outside the box' has become the most ridiculed of management clichés for a reason.

The Blind Planet

In the heat of Britain's 2016 Brexit referendum, one of the country's most senior politicians, Michael Gove, made a surprising claim: 'people in [Britain] have had enough of experts'. Surely the modern world relies on experts; when you get a cancer diagnosis, fly an aircraft or evaluate complex legislation, experts are helpful? Not any more! For a substantial portion of the population, Gove's comment rang true. In the referendum most experts were duly ignored by a populace comfortable with Gove's description. Across the Atlantic, the message, and the impact, was the same: 'The experts are terrible,' said then presidential candidate Donald Trump.

Just as we had reached new peaks of complexity and difficulty and required expertise more than ever, the very notion had become toxic, the emblem of a self-interested and pompous elite; the problem, not the solution.[74] This was hardly the best preparation for a global pandemic.

A barrage of populisms means that ideas are adversely re-politicised. Towards the end of the twentieth century, the exploration of new ideas had never felt less controversial. This now looks like a chimerical interlude. While populism is typically taken to be the wave of broadly nationalist sentiments that has surged since the 2010s, I take it here in the widest sense, as shorthand for a spectrum of intersecting social pressures. From demagogic leaders to grassroots backlashes, to a vicious intensity of partisan debate, to a focus on hype and headlines, populism is in a deep and global intensification. Although contested, it is protean and ineluctable, and is infiltrating into every corner of endeavour. The coming decades will unfold amid the fault lines and fractures of the populist surge.

Experts face a crisis of authority. Large numbers of people

no longer believe doctors, or journalists, or legislators.[75] Trust in government and the media are at record lows around the world: four out of five Americans, for example, believe Congress is corrupt.[76] But this lack of trust extends to institutions and expertise in general. Confidence in the medical, scientific and educational establishment has plummeted since the 1970s and is still falling.[77] Even within a disciplinary niche such as social science, a survey of practitioners found that trust in their own area was falling.[78] The Edelman annual trust barometer records that across many nations institutional trust is at an all-time low. The people and institutions who can and should foster new ideas have never been so unpopular.

Polarisation is growing, spaces of common ground are depleted. Increased partisanship closes minds, lines of enquiry and potential collaborations. Heterodox opinions, processes and people cannot find purchase in environments of ideological purity. In the 1990s there were around two left-leaning professors for every one on the right. By 2011, this was five to one.[79] In the arts, humanities and social sciences the ratio is well over ten to one and often much higher.[80] Even economics, the most right-wing of all the social sciences, has a four to one left-leaning ratio.[81]

Over the period academia became better (if still wanting) in its understanding of gender and racial diversity, but homogenised in terms of its core ideas and worldview. In many ways I share the political biases of the contemporary academic establishment. But I do question whether it doesn't have downsides and is symptomatic of a more fundamental groupthink. Such cognitive homogeneity, the in-group flip side to polarisation, is likely reflected in the corporate world as well. Splits in academia echo splits in society: in the US 81 per cent of those identifying with a given party have a negative opinion of those on the other side, a high.[82] Partisanship, defined as how easily a member of Congress' political affiliation can be identified from a single sentence, was flat from the late nineteenth century to the 1990s. But it increased steeply from the 2000s, rising from 55 per cent to 83 per cent of sentences.[83]

Tribalism is resurgent, fuelled by online echo chambers that entrench partisan agendas on everything from local politics to medicine or child rearing. How can different ideas flourish amid a more intolerant world, both nationally and internationally? Different parties, groups and countries cannot even agree on the problems. So-called epistemic polarisation even impacts those who can agree on end goals – simply regarding evidence from others as uncertain can begin a process of escalating polarisation and distrust.[84]

This reflects a wider closing of minds. In the 1960s students rioted in the name of academic pluralism and free speech. Today they riot to no-platform speakers they disagree with. Old-fashioned values about truth seeking have been compromised in favour of ideologically driven positions. Safe discourses and spaces have the potential to stifle debates. Consensus and a safety-first approach to the intellectual sphere are non-negotiables for large portions of the student body and the commentariat alike.

Too often people either walk on eggshells or are deliberately provocative. Tactics which might have been legitimate for shutting down trolls and extremists have infiltrated the entire discourse of ideas. Callout and cancel cultures breed self-censorship. Oppositional us-versus-them culture wars shut down conversations and erode any good-faith sense of a collective enterprise. This is an atmosphere that closes down 'free inquiry, dissent, evidence-based argument and intellectual honesty'.[85] Make the wrong comment and you are publicly shamed. Take the wrong view and you are no-platformed.[86] It all dampens appetites for risky, brave, original thinking or comment, a process that flows down to topics of study, methods, what counts as an important question or answer. It also delimits what matters are available for debate. Some concepts and narratives are placed beyond discussion, and woe betide the individual who strays, even accidentally. Disciplines curdle into self-confirmatory bubbles. I have seen this happen to literary studies, much of it becoming pious, inward-looking and intellectually unambitious. A puritan orthodoxy has pervaded the entire field.

Portions of the right are egregiously bad, flinging around absurd claims that 'cultural Marxists' or postmodernists are at the root of all the world's problems. Many on the populist right have abandoned reasoned argument and evidential standards altogether, spearheading the assault against expertise and so-called elitism. In the US an extreme right wing has developed a new establishment anti-science, anti-knowledge. The Republican Party's unholy alliance with the religious right is an odd symptom of this, especially given that Founding Fathers like Thomas Jefferson, Benjamin Franklin and John Adams had serious commitments to scientific standards, knowledge and debate.[87] A phenomenon that first flourished under Reagan, who supported the teaching of creationism in the 1980s, it has continued and intensified under George W. Bush and Trump. *Nature* argues it could take decades to recover from the damage to public trust in scientific enterprises brought about by this populist onslaught as medical advice is ignored, climate change dismissed, entry routes for new scientists cut off and scorn poured on the measured institutional bedrock of truth-seeking, collaboration and deliberation.[88]

This in turn gives succour to wider currents of anti-science and anti-reason. Diverse and prevalent examples include the ongoing role of pseudoscience in 'alternative medicine' like homeopathy, anti-vaccine propaganda or even the quack remedies of celebrity wellness sites; anti-evolutionary/creationist beliefs; the denial or downplaying of climate change; the promotion of paranoid conspiracy theories, whether local or international; national or race-based sciences and knowledges; the warping, traducing and making up of social scientific data and arguments in the name of political positions; the dismissal of anything from stem cell research to genetically modified crops on the basis of shoddy arguments and paranoia; and the specious posturing of a cadre of so-called intellectuals from all angles of the political spectrum. Ideas of what might constitute theory or evidence or proof are all set spinning: each can simply be what anyone wants it to be, a curious meeting of radical postmodern theory and the online alt-right media.

Nowhere was this clearer than during the pandemic. Government agencies like the Centers for Disease Control and the FDA in the US were subject to extreme political pressure. From Wuhan to Washington, Brasilia to Brussels, ideas responding to an immense challenge were politicised, becoming not so much a space of exploration as a battleground on which to destroy one's enemies. Areas as diverse as policy, epidemiology and even biology became subject to the criterion of prior political views. Questions like whether you should work at home or wear a mask were no longer seen as problems to be solved, but badges of allegiance. In general, Covid-19 was a Petri dish for many of populism's suffocating ideas: extreme tribal mindsets; the ascendancy of opportunism over substance; reams of misinformation; collapsing standards of discourse and research.

I don't like wading into politics. I want simply to note that the intellectual ferment stews in these problems; that there is no sign of this letting up; and that it prioritises forms of discussion that close down possibilities for big and original ideas. It's also not so much each individual debate that is worrying but the wider force: the spectre of a post-truth world where anti-intellectualism and in-group conformity triumphs. This has a long history of course, but is given a second bite of the cherry in a world where freedom of thought had supposedly won.

Campus culture or social media spats are the thin end of the wedge. A return, politically and culturally, to global nationalism and nativism turns vast populations inwards and backwards. Both the American and the Australian government (let alone elsewhere) have suggested that federal money should only be spent on research or cultural projects that are, in some sense, 'patriotic'.[89] The Brazilian government puts a creationist in charge of its science policy. In India, academic texts the government deems anti-nationalist are expunged from curricula.

The rise of ethno-nationalism in America, Turkey, Russia, India, China – almost everywhere, in fact – spurs techno-nationalism, which in turn threatens to splinter the common standards necessary for scalable innovation. Cross-border collaborations, on films or businesses or research, become less likely. Tariffs diminish trade. Cultural exports are especially limited. Horizons shrink. A dangerous conflagration becomes more likely. A chest-beating nationalism suffocates the twenty-first century Republic of Letters. As nation states cement their place in the geopolitical firmament, they are unlikely to answer the kind of transnational questions discussed in this book.

And over the long course of the century, fundamentalisms of different stripes will continue to flourish. Demographic trends favour the most extreme forms of religion against more secular groups: the more extreme your religious views, the more children you have.[90] If history really is demography, it doesn't bode well.

Meanwhile all sorts of unintended consequences can impede ideas. For example, Amazon's R&D budget is over four times the size of the entire UK government's research expenditure. Yet tech companies, the last bastion of free-wheeling innovation, have left their salad days behind. Not only were they spending, but there was also a sense that these were the last organisations, with their labs and moonshots, willing to tackle grand challenges with a can-do mindset and fresh thinking; the last hopes for Bell Labs 2.0. Now tech faces significant (and arguably self-inflicted) obstacles: accused of violating social norms, monopolising sectors and fomenting a series of populisms, companies face a 'techlash', both among users and in the regulatory capitals of Washington and Brussels. Outsourcing radical innovation and thinking to a handful of West Coast businesses with lots of money was always unwise – and yet it is the unthinking strategy for much of the Western world.

If economic conditions warp the ecosystem of innovation, social pressures have a similar impact on the ecosystem of abstract ideas. The political scientist Daniel Drezner argues that the public sphere has become an 'ideas industry': commodified, politicised, chiselled down, neatly packaged to capture that currency of the modern world, attention.[91] Research groups, startups and government departments are all chasing the eye-catching press release, the wave of hype that will carry them to glory. Screw evidential standards and slow, messy research: bask in the glory of a breathless headline, a different but no less insidious brand of populism. We move from 'public intellectuals' to 'thought leaders'; from critics and sceptics to evangelists; from open to closed minds; from expertise towards personal experience, however shallowly constituted; from lecture series to ten-minute TED talks; from books to blog posts; from scholarship to the consultancy gig;

from disinterestedness to the impact agenda; from thoughtful correspondence to Twitter; from research for research's sake to research for plutocrats and autocrats.

Discussion is moving away from a commitment to academic, critical, inquisitive values towards the more deliberately provocative and point-scoring.[92] It has business analogues in a commercial system primed to offer pointless services and quickly discarded gadgets; cultural analogues in the endless production of same-same eyeball-sucking pabula. It's founded on institutional erosion: where nominally free places of enquiry, like think tanks, become corporate-funded and ideologically driven attack dogs, or where museums and art galleries self-censor in the race for sponsorship. Perhaps it is also founded on some as yet unquantified mental erosion: the constant busyness and distraction of our blinking social media notifications, chipping second by second at our capacity for deep thought.

This is the post-truth age, one of 'alternative facts', tribal loyalties, quick bucks and information warfare. In the words of another political scientist, 'feeling took over the world', whether that's distrust of experts or faith in the nation above all.[93] It is an age seeing a turn towards atavism, the rise of an 'identity-protective cognition' where people's ideas are based on who they are, the erosion of reasonable discussion – even, in some places, of reason. Special interest groups capture the public conversation. Forces of xenophobia overwhelm those of collaboration; forces of extremism those of curiosity. An idea's political, populist import has become, at times, the cardinal criterion of value or significance. Confidence in proposing and experimenting with new ideas is damaged as a result. So much for the notion of a post-Enlightenment world where ideas or policies or theories or artworks are judged on merit.

This is about our tolerance; about openness to the strange

and challenging; about a societal willingness to seek and confront new evidence and opinion, to think clearly and think again. And about how, brick by brick, that capacity is being chipped away.

Society has never been kind to big thinkers. Ask Socrates. Ask Jesus. Ask the astronomer Giordano Bruno, who was hung upside down naked before being burned at the stake for his heretical – what we would now call scientific – ideas. Ignaz Semmelweis was crushed by lack of recognition for his theories about hand washing, an innovation which would eventually save millions of lives – put in an asylum, he was beaten by guards and died from an infection. During the heat of the Second Industrial Revolution the Italian Ministry of Post and Telegraphs suggested Guglielemo Marconi be referred to an insane asylum.[94] Even in Vienna's crucible of thought, Freud was ridiculed and dismissed. For most of recorded history, breaking artistic 'rules' wasn't cool; it was monstrous. And compared to food, security, land, power, sex and riches, ideas were never high on the agenda.

As we have seen, there has on the surface never been a better time for breakthroughs than today. However, just because we compare well historically does not mean we have reached some ideal situation. All kinds of pathologies and problems persist in the suffocation of radical ideas, excreted from every pore of the contemporary world, too large, too dispersed, too interwoven to be dislodged. Expanding megacorps with return-hungry webs of ownership; ageing populations; regulatory accumulation; a frenzied atmosphere of bitter polarisation: only a brave forecaster would predict these will be rolled back. Even the world's most powerful entities – the EU, the US federal government, the Chinese Communist Party, the tech giants – cannot contain or direct them. And all of this represents a business-as-usual approach to the future – calamities from an even more extreme pandemic, war, state failure and so on, could yet make things much more difficult still.

Every society presses down on big ideas in a variety of ways, but also creates pockets of opportunity: institutions, individuals, mechanisms, funds, discourses, publications, trends that allow for revolutionary ideas. The question is always how big or small those pockets are drawn. Bell Labs was such a pocket. Vienna overall was staid, but its tensions created a remarkable cluster of pockets. We've drawn our pockets tighter than we could and should. It's not then that we're bad at big ideas; it's that powerful, diffuse forces of stagnation have quietly accumulated even as older problems disappeared.

Think of medical progress, and how society exacerbates the breakthrough problem here. It is subject to the Idea Paradox. Easier-to-treat conditions, obvious compounds and drugs, clear public policy health wins: they've been taken. What's left are more complex and intractable diseases like Alzheimer's, or new public health crises like the opioid epidemic or small particulate pollution.

But societal context doesn't help. A combination of perverse incentives, flawed models and burdensome (but in many cases irreplaceable) bureaucracy creates friction. Incentives for drug discovery are skewed: companies are disincentivised from targeting conditions that will, thanks to their intervention, quickly clear up. Private money is channelled into where it can deliver returns to dividend-hungry shareholders. The London-based Office of Health Economics estimates the net present value of a new antibiotic at just a twentieth of a chronic neuromuscular drug.[95] Antibiotics solve acute problems; but chronic conditions are more profitable. Where research is difficult, in brain disorders for example, Big Pharma has recently scaled back its efforts or pulled out entirely.[96] Regulations are colossal, meaning startups' only chance of getting to market is via acquisition.

There are also much stricter ethical safeguards than in the golden age. It took only hours to create the BioNTech Covid-19 vaccine, but months to see it through trials and regulatory approval. In the late 1960s, liver transplantation was seen as a triumph, but only two of the first thirteen recipients survived the process.[97] If researchers published a mortality rate like that today, they would, at the very least, find their research paused. The ever-escalating cost and complexity of clinical trials is the byproduct, trials that now only Big Pharma can contemplate tackling.

A Tufts University study estimates that the length of clinical trials has increased by 70 per cent, and that 67 per cent more clinical staff are needed to administer them.[98] Some question the direction of the underlying research: mega-budgets, careerism and corporate organisation don't allow for the maverick and serendipitous approaches that have often worked well in the past.

The same nexus underwrites the breakthrough problem in transport. The engineering challenges here are immense. Going to Mars is much more complex than going to the Moon. We don't yet have reliable, reusable heavy-lift rockets, let alone the infrastructure needed to sustain lengthy deep-space cruises and closed-loop permanent settlements. While ambitions for space exploration are escalating, in practice budgets are not. Even the rockets for Artemis, the next planned Moon missions, are billions over budget.[99] Solving the problem of autonomy in cars has been hailed as a massive leap. But despite undoubted progress, one prominent investor puts us at least $40 billion of single-company R&D away from fully autonomous vehicles.[100]

There are stiff barriers to gaining purchase. All forms of transport are subject to extensive regulation and new forms often require new infrastructure, at eye-watering cost and inconvenience. Think about flying cars: even when the idea is plausible, do city residents want loud swarms of them buzzing overhead? Who will pay for their landing sites? What happens after the first crashes in which bystanders are killed? What price will insurers charge? Such barriers are not insuperable, but they are exacting and omnipresent compared with the early days of the car or the plane: Stephenson, Benz and the Wrights didn't worry (so much) about insurance, or licensing, or safety regulations, or litigation, or infrastructure, or impatient venture capitalists, although they might have had to grapple with a sceptical public, protectionist governments or a hostile media. As J. Storrs Hall points out, if Henry Ford had been sued every time someone crashed their Model T, we likely wouldn't have cars and highways at all.[101]

New forms of transport or medicine bump up against tight economic and policy limits. Concorde ceased flying not just over safety fears, but because it made a loss. It's possible to build hyperloop trains, hypersonic aircraft or lunar-orbiting bases, or send a mission to Mars, but the delicate balance of risk, regulation and return on investment may make them infeasible. Financiers won't fund new services for which they believe there will be a lack of demand, or that are likely to hit regulatory headwinds, and because the engineering is so challenging, two brothers in a workshop can't help us this time. Ideas find themselves caught in a vicious loop, stifled before they can reach execution, stuck in the heads of a few wild-eyed entrepreneurs.

Nowhere are the effects of the stagnant society more apparent and deleterious than in the fight against climate change. The facts here are no secret. Even at two degrees of warming, the very lowest possibility, we face the prospect that there will be hundreds of millions of climate refugees by 2050.[102] Deadly heat will become common. Catastrophic sea level rises will flood some of the world's most densely populated areas, from New York to Shanghai. Occurrences of extreme weather and flooding will increase, even as wildfires proliferate and conflict surges. Potable water and agricultural yields will collapse. As a recent *Nature* report argues, we may already be at the point of 'a global cascade of tipping points that led to a new, less habitable, "hothouse" climate state'.[103]

We know what we need to do: make a 45 per cent reduction in CO_2 emissions by 2030 and reach net zero by 2050. But modern society is soaked in carbon emissions: the food we eat, the water we drink, the clothes we wear, the incidental details of our day-to-day lives. In the words of David Wallace-Wells:

> The scale of technological transformation required dwarfs any achievement that has emerged from Silicon Valley – in fact dwarfs every technological revolution ever engineered in human history, including electricity and telecommunications and even the invention of agriculture ten thousand years ago. It dwarfs them by definition, because it contains all of them – every single one needs to be replaced at the root, since every single one breathes on carbon, like a ventilator.[104]

Another way of thinking about this is that we would need to add the equivalent of a nuclear power plant to the global grid every single day for the fifty years from 2000 to 2050. Renewables have become much cheaper, yes, but still coal use is way up in the new millennium. The revolution is on track, but at this rate we'll get there in 400 years.[105] Nowhere is the demand that we step up and deliver big new ideas more urgent than in the matter of planetary survival. There are a number of ideas that could work. A massive carbon tax. Radical change in consumption patterns in everything from travel to diet. Huge investment and rollout in carbon capture and storage. New pushes in agriculture and forestry. Blue sky technologies from fusion to, potentially, geoengineering. There is so much scope: just a few hours of the sunlight that falls on Earth would give us more energy than we need today; 20 per cent of global wind capacity is seven times more than we need.[106] We've already looked at fusion power. We can regear our approaches to land, agriculture and the food supply chain; transform our materials superstructure and manufacturing base. Create genuinely sustainable modes of living and consuming. Build carbon neutral or, better yet, negative technologies and economies.

In the face of all this, perhaps the ultimate problem is our inability to pursue collective action. Bound by the needs of companies or nations, we're hemmed in by a context of short-term financial interests; a parochial lack of vision; political and national tribes still looking out for their self-interest. The fundamental truth of global society is that the planet is burning and yet none of our efforts at the frontier have so far measured up. Exhibit A of our stagnant world is one incapable of rolling out ideas to save itself.

You might think that, for example, the founding of a new musical genre is far removed from the climate crisis. In one sense it is, but in another it's about dynamism, attitudes to trying and delivering new things at the frontier. Its absence is indicative of a deep complacency. Tyler Cowen has gathered much evidence of this: people don't move house but stay in the same jobs and social class, don't even leave the house as much as they used to.[107] Segregation, matching, risk dodging: everyone works hard to avoid new and challenging cultures and experiences. Resistance to change has always been a feature of life, but after decades of greater openness, we have now reverted to type. A civilisation addicted to complacency is not one well disposed to meet the colossal challenges of its times.

These two chapters paint a grim story for the future of big ideas, suggesting a continuation or even an intensification of the trends described in Part I. J. Storrs Hall talks about ideas as an onrushing tide over a landscape.[108] Occasionally it finds a new fertile valley and rushes to fill it. This is a time of paradigm shifts and phase transitions, of frontiers pushed back. But eventually the valley is filled. Ideas stabilise, the valley becomes a lake. As Storrs Hall argues, the stagnant society has given up on torrents and dammed its current valley. Just as we should be building aqueducts to more distant valleys, we're content in our lake.

But it's not the end of the story. There are still pockets, signs of new aqueducts tentatively under construction, glimmers of a major push towards far and inaccessible valleys. Ideas are thus caught in a tension. We've seen the negative pressures that explain many features we saw in Part I, from the breakthrough problem to diminishing returns. It's time we looked at the other half of the equation, the reasons to take heart.

7

The World's New Toolkit

Artificial Ideas

The protein-folding problem is one of the most significant and intractable in contemporary science. It works like this.

Proteins are the building blocks of life, responsible for everything from the movement of limbs to the detection of light; they form the basis of the body's tissues, our muscles, bones and hair. Seventy-five per cent of dry body weight is made up of 100,000 different kinds of protein.[1] Compiled from chains of amino acids, they first form sheets or helices, and then fold into elaborate three-dimensional structures. A protein's intricately folded shape defines its function. Antibody proteins form hooks that latch on to viruses and bacteria, marking them for attack by the immune system. Proteins that form elements like cartilage and skin – collagen proteins – resemble cords.

Amino acids are programmed by DNA. But only the sequence is programmed. How they actually fold is an unsolved problem; DNA prescribes what shape might be formed but doesn't let you automatically guess. The upshot is that, despite massive advances in DNA sequencing, you cannot easily predict the 3D structure of a protein based on a genetic sequence.

There has been steady progress. X-ray crystallography, nuclear magnetic resonance and cryo-electron microscopy have all helped to uncover protein shapes. Powerful algorithms have made a great difference, but there is no way to get an answer by brute force. It would take 13.8 billion years to calculate all the different ways a protein could fold – longer than the age of the universe.[2] Yet they do fold, almost instantaneously.

If we are to understand life, we need to better understand the relationship between DNA and protein folding. Doing so would profoundly improve our understanding of our own bodies and of biology in general. Only around half of the proteins in human cells have ever been mapped. Improving this ratio means we could learn more about rare diseases, understand how they arise from small changes in DNA and, based on this knowledge, more efficiently design drugs to help. Indeed, the whole process of drug discovery could be supercharged. But there's more: understanding protein folding would help us design proteins. We could create new organisms to help combat climate change or break down rubbish and waste plastic.[3]

Over recent decades research groups have chipped away at the problem. To keep score they settled on a competition, a global methodological tournament – the biannual Critical Assessment of Protein Structure Prediction (CASP).[4] Here rival teams pitted their work against one another, the winner being decided by whoever had the best predictions of how a given protein would fold. This was a spur to progress; a Petri dish for trialling different techniques and approaches, all benchmarked in the most public and competitive fashion. Biology's toughest nut had gamified scientists' competitive instincts. Begun in 1994, it quickly became the litmus test of progress in a tight field. It had attracted attention in the 2000s for a new kind of outsourced science: at the University of Washington in 2007, software called Foldit sent the protein-folding problem further into the world of gaming. Researchers created an online video game where fun and fiendish puzzles happened to advance science. Players, who could participate from anywhere in the world, were scored by creating low energy shapes, which in turn offered scientists clues as to the likely shapes of folded proteins.

At CASP 2008 and 2010 Foldit was competitive with the world's top teams, but it had reached diminishing returns. Then at CASP13, held in Cancun in 2018, a new entrant took the field by storm. Having no record or background in the field, they beat off ninety-eight competitors to record a decisive win. According to the organisers, they made 'unprecedented progress'. Mohammed AlQuraishi of the Harvard Medical School said the dominant mood at Cancun was surprise. People were asking 'What just happened?'[5] There was a sense of 'melancholy', 'existential angst', at how it was possible for outsiders to make a jump that, in AlQuraishi's words, worked at twice the pace of regular advance, and possibly more. It was 'an anomalous leap' in one of the core scientific problems of our time.

What *did* just happen? The artificial intelligence company DeepMind, part of the Alphabet group, had been quietly working on software called AlphaFold. DeepMind uses deep learning neural networks, a newly potent technique of machine learning (ML), to predict how proteins fold. These networks aim to mimic the functioning of the human brain, using layers of mathematical functions that can, by changing their weightings, appear to learn. This makes predictions which are scored against what is already known. Based on its results the neural network keeps learning, getting more accurate the more it guesses to build a highly effective model. At

CASP13, AlphaFold was better able to guess the distance and angles between pairs of amino acids. It was a simple, unexpected but powerful approach, supported by world-leading engineering and ML.

DeepMind didn't invent this approach, but they pushed it harder and further than anyone else. They coupled scientific insight with awesome engineering power. Unlike pharmaceutical companies they didn't have a stake in the space, yet they almost immediately leapfrogged the incumbents. AlQuraishi castigated Big Pharma: how could they, so wealthy and so directly implicated, let outsiders wipe the floor with them? It was 'laughable' that they had neglected basic research to such an extent. Despite hesitations, the academics were ultimately positive. Progress was progress. In the words of Paul Bates of the Francis Crick Institute, 'They did blow the field apart.'[6]

The truth is, as the Idea Paradox suggests, that we are left with the truly hard problems; and protein folding is a problem of savage complexity. It demands constant ascent through the technological and methodological gears. Had a new gear been found?

DeepMind was already known for its ambitious use of ML. Founded in London in 2010, its stated goal was to 'solve intelligence' by pioneering the fusion and furtherance of modern ML techniques and neuroscience: to build not just artificial intelligence (AI), but artificial general intelligence (AGI), a multi-purpose learning engine analogous to the human mind. DeepMind made headlines when it created the first software to beat a human champion at Go. In 2016 its AlphaGo program played 9th dan Go professional Lee Sedol over five matches in Seoul and, in a shock result beyond even that of CASP13, won four of them. This was years, even decades ahead of what anyone had expected. There are 10^{82} atoms in the observable universe but 10^{172} possible positions in Go;

this makes it exceptionally difficult for classic machine-driven approaches, exponentially tougher than chess.[7] Only a new approach to AI could have triumphed.

Sedol moreover was an intuitive, virtuoso grandmaster, a difficult opponent by any measure. The pivotal moment in the contest was the now-legendary thirty-seventh move of the second game. AlphaGo played a move completely outside the game's conventional thinking. It just didn't make sense. But later in the match it proved decisive. With that move Go was changed forever. So was our worldview; machines could forge new paths, paths hidden from us. They could be radically creative and deploy revolutionary insight. In the history of Go, move thirty-seven was a big idea that, in thousands of years, humans hadn't thought of. A machine did. Thanks to the program, previously unthinkable moves are now part of the tactical lexicon. AlphaGo, like AlphaFold, jolted the game out of a local maximum.

DeepMind is at the forefront of a well-publicised renaissance in AI. (AI itself is a big idea that goes back to Alan Turing and pioneers like John von Neumann and Marvin Minsky and, in the form of dreams of automata, much earlier still.) Over recent decades, computer scientists have brought together a new generation of techniques: evolutionary algorithms, reinforcement learning, deep neural networks and backpropagation, adversarial networks, logistic regression, decision trees and Bayesian networks, among others. Parallel processing chips have boosted computational capacity. Machine learning needs vast amounts of 'training' data: these technical advances have come just as big datasets exploded. Business and government piled investment into R&D. Rapid improvements in areas like image recognition, natural language processing, translation, game playing and autonomous driving transformed services and generated hyperbolic headlines. The history of AI is one of crests of hype followed by troughs – the so-called AI winter set in from the 1980s with the broad failure of 'symbolic' approaches. But by the 2010s a new spring had arrived, and DeepMind was in the vanguard.

Public conversation around AI has been dominated by the risks and rewards of job automation. And yes, this is a significant question. Nonetheless, I have yet to meet an AI scientist motivated by the prospect of automating a call centre. Instead, they are motivated by the prospect of discovery and knowledge far beyond our present abilities. AI scientists are nerds; above all they care about science and ideas. Yet what AI will do to human knowledge, to our ability to comprehend and see and discover and create, has received coverage incommensurate with its potential impact. That should change.

AI is a cognitive technology, a meta-idea, and so goes to the heart of questions about how ideas are produced. New forms of knowledge and perception, quite unlike those of humans, beyond our unaided capabilities, are starting to accelerate the production of ideas. AlphaGo and AlphaFold are signposts to an era where those closest to a particular toolset are best positioned to push back the frontiers of knowledge. Proximity to these tools helps accelerate discovery, producing watershed moments like those at Seoul and Cancun, not to mention other moves from DeepMind alone into areas like medical diagnosis and the modelling of physical processes. And as move thirty-seven hints, this isn't just a matter of inching forward or copying humans; it adds a qualitatively different dimension.

If that sounds overdone, consider that AlphaFold wasn't finished in Cancun. Two years later, as I was writing this book, at CASP14 a new program, AlphaFold2, emerged. It performed even better, so well in fact that some claimed the problem of protein folding had finally been solved, a shocking leap in progress on even the first program. The headlines were breathless: *Nature* said 'It will change everything', *Science* that 'the game has changed'.[8] If anything AlQuraishi was even more stunned than before, blogging that the problem he'd spent his life trying to solve was now potentially done, leaving him feeling like a child had left the family home.[9] 'It's a breakthrough of the first order, certainly one of the most significant scientific results of my lifetime,' he said. A fifty-year grand scientific challenge was over, a new realm of biological possibility and understanding broached by the latest AI techniques. AI is already re-accelerating scientific discovery.

Demis Hassabis, the co-founder and CEO of DeepMind, talks about this aspect of AI. As he argues, 'The promise of AI is that it could serve as an extension of our minds and become a meta-solution,' and in doing so help 'usher in a new renaissance of discovery, acting as a multiplier for human ingenuity, opening up entirely new areas of inquiry'.[10]

Our tools, and the potential they embody, are not standing still. Those longer fishing rods may be coming.

Tools aren't neutral in the processes of thinking and creating. They're part of what we imagine and do; not just means to ends, but ways of seeing, modes of discovery. New tools represent big ideas in themselves. But they are also platforms, catalysts, amplifiers and enablers of further ideas. Radical new tools are the most likely means by which we can leapfrog problems and uncork bottlenecks around big ideas. The prospect of an upgraded underlying toolkit is thus of immense significance.

Consider the Scientific Revolution of Copernicus, Brahe and Kepler, the birth of modern science, arguably the most important moment in human existence since the Neolithic Revolution.[11] At its heart were potent conceptual tools that introduced a self-sustaining process of ideation: the creation of concepts like 'discovery', 'fact', 'experiment', 'law of nature', 'hypothesis' and 'theory' established the mental architecture for what eventually became known as the scientific method.[12] Without this framework, nothing else about modern science – and arguably the modern world – would make much sense. Important too was a set of intellectual tools like calculus, Cartesian graphs, algebra and probabilistic mathematics.

But the revolution was also galvanised by novel instruments. In the medieval period writing was generally on vellum: calf skin. It was expensive. The advent of rag-based paper meant lower costs – information could spread more easily. And this was as nothing to a later innovation, the printing press. It allowed the inclusion of detailed, accurate images alongside the text. Its greater productivity knitted dispersed researchers into a community. Works were disseminated faster and more widely – giving them not only more reach but better preservation in the public sphere. Words, data and images were fixed across a print run, letting readers check results and experiment against them to create common areas of understanding, standardised 'facts'. The printing press was vital in orchestrating hallmarks of scientific enquiry like organised scrutiny and replication.

That other great revolution of the sixteenth century, the Reformation, also turned on the functionality of the press. Whereas previous revolts against the Catholic Church had faltered, the Europe-wide dissemination of Luther's message saw it reach critical mass (if you'll excuse the pun).

Printing spurred literacy, which spurred learning. It introduced the world to mechanised, industrial workflows. The press was just a tool, built like every tool from other tools and techniques (such as the winepress and metallurgy in this case), but its impact was of universal importance. As the signature invention and enabler of that pivotal epoch we call the Renaissance, it reflects the key tools of our time: a historic big idea in itself, but also a platform for new big ideas. Not just a passive item but, in the words of its great historian Elizabeth Eisenstein, 'an agent of change'.[13]

Optical devices were also central to the Scientific Revolution. Galileo's initial innovation was to improve telescopes, instruments originally patented in the Netherlands. Where he lived in Padua, then part of the Venetian Republic, these had an obvious naval use. In 1609 a demonstration to the Senate from the tower of St Mark's was a notable success. But Galileo's were too powerful for his city masters – his were adapted for the heavens more than the Lagoon. By 1610, having ground hundreds of lenses, he achieved a 30× magnification, much more than contemporaries who made do with 6–8×.[14]

Using these lenses, Galileo made a series of stunning observations that buried the old Ptolemaic universe once and for all. He discovered the moons of Jupiter, astounding Europe, watched the phases of Venus, proving that it orbited the Sun, saw Saturn's rings, and observed sunspots, evidence of change in a previously fixed realm.[15] The Milky Way was, incredibly, shown to be 'a mass of innumerable stars planted together in clusters'.[16] Telescopes introduced a kind of space travel, changing perceptions of the solar system, space, the place of the Earth and humanity. It suggested that there might be other worlds like our own, inhabited, vibrant, and it hinted at a universe much larger than suspected. It is hard to think of bigger ideas than these.

Later, in the Netherlands, Antonie van Leeuwenhoek looked through powerful single-lens microscopes with up to 500× magnification. In 1676, peering through a focused glass bead, he saw living creatures invisible to the eye, another astonishing order of existence. While it took only months to confirm Galileo's observations of the moons of Jupiter, such was the level of magnification that it was four years before the existence of Leeuwenhoek's minute organisms was confirmed.

A world as intricate and complex as that on the human scale was revealed. The microbiology that Pasteur did so much to advance was conceived only with the help of this new instrument. It was as revolutionary a moment as Copernicus and Galileo, upturning assumptions about who the world was for and how it worked.

Telescope and microscope are vital to what historian David Wootton calls the 'Scaling Revolution' of circa 1610–1700: one of the most significant events in intellectual history, where the size and scope of the world was drastically expanded both up and down the scale.[17] It showed the limits of our sensory organs, and implied that however large or complex you thought the world was, you should think again. Even more than Copernicus, the Scaling Revolution destroyed the old, static cosmos; after all, for Copernicus the solar system was still all there was, just the Earth orbiting the Sun, both of them unique in the universe. All that began to crumble. Shown here was a universe vast and cold and beyond us.

Other tools made a difference. There was Robert Boyle's air pump, Newton's prism, Fahrenheit's thermometer. What was invisible became measurable, and what was measurable could be known. New techniques of compass reading, map and ship making were vital in reaching the 'New World'. The development and application of mathematics spurred engineering – as with Brunelleschi's dome in Florence. Biology required new techniques and attitudes towards dissection. Clockwork mechanisms refined the measurement of time and spurred, again, a new approach to engineering.

The Scientific Revolution wasn't simply a revolution of ideas. It was a revolution of tools. It laid a lasting pattern: the underlying toolset of an era is not a passive product of its science or economy, but a co-creator. As we have seen throughout history, and contrary to the popular myth that everything flows down from ivory towers, tinkerers and tools often open the spaces for new insights.[18] The invention of the steam engine preceded the understanding of the laws of thermodynamics upon which it relies. Vaccination preceded the knowledge of antibodies. Pasteur was as much motivated by the practical concerns of wine growers, silk makers and the sick as by the imperatives of pure science. Films require a camera, the computer game a computer.

Without its toolkit, the Scientific Revolution – not to mention the Reformation and Renaissance – would look very different. This is true of everything since – tools and ideas work in tandem.[19] In the twenty-first century our capacity to develop big ideas will rest on the development of our tools more than any other factor. Hence the significance of AI. It is the calculus, the telescope, the compass of our time.

Demis Hassabis himself makes the link explicit, calling AI a sort of general-purpose Hubble space telescope for science.[20] Big ideas like AlphaFold and AlphaGo, instances of the big idea of deep learning neural networks, are steadily making a difference at the coalface.

To see how AI reshapes ideas, consider the volume of data produced by contemporary experiments. At CERN, the Large Hadron Collider produces 25 gigabytes of data every second.[21] NASA missions churn out more than 125 gigabytes of data every second.[22] Climate scientists, particle physicists, population ecologists, derivatives traders and economic forecasters all generate and must process vast amounts of data. Each moreover addresses complex dynamical systems. Advances must reckon with this mutating tsunami of data to stay current. No human or humans could ever work at this scale or with this level of detail unaided.

Already ML is integral to each of the above cases, curating

and processing the deluge. It is making a tangible dent in the breakthrough problem. The authors of the original Eroom's Law paper now believe the era of stagnation may be coming to an end thanks to the prevalence and new-found effectiveness of machine learning in the discovery of drugs.[23] AI is moving to the front lines of the battle against cancer and a paper in *Cell* illustrates that ML can use molecular structure to predict the effectiveness of antibacterials (the researchers behind the AI even called the resulting antibacterial 'halicin' after HAL, the AI in *2001: A Space Odyssey*).[24] We need things like this to beat future pandemics. Fusion scientists are optimistic that the application of AI could bring decisive advances in the coming years, and in general the field is now focused on ML approaches to core problems.[25]

Breakthroughs in natural language processing are coming at pace: the parameters of OpenAI's eye-catching GPT language prediction system grew from hundreds of millions to hundreds of billions in just a few years with some spectacular results, enabling it to write convincing text at length on any subject.[26] GPT-3 can take a portion of writing and then continue it with at times shocking plausibility. This is a powerful real-world application already throwing up startling ideas. If you have a chance to play with such generators, nothing so immediately conveys the speed and potential of this new age.

As sophistication and computational capacity grow, so does our ability to see new things in the data, beyond the limits of human perception. To give a concrete example, take materials science, a subject which matters not just for our curiosity but for meeting serious challenges: help with materials could spur on fusion, or revolutionise battery or photovoltaic technologies.

A team at California's Lawrence Berkeley National Laboratory gathered 3.3 million abstracts from published papers on materials science from over 1000 journals between

1922 and 2018.[27] They then fed these into an algorithm called Word2vec which analysed the relationships between words. The results were impressive: the algorithm could predict thermo-electric materials way in advance. By stopping the analysis some years ago they could compare the algorithms' predictions against what actually happened, verifying their validity. It's concrete evidence that the algorithm could have fast-tracked scientific discovery. From abstracts alone, the algorithm had taught itself the structure of the periodic table and the nature of crystals. And, what's more, it was predicting new, improved and as yet unknown materials. The team were genuinely surprised.

The significance is twofold. First it suggests a way out of the burden of knowledge problem. This is a system specifically designed to cut through a mass of learning impossible for any human to digest. As knowledge accumulates, so technology like Word2vec can sift through it. Second, it shows how even at a comparatively early stage, AI has enormous potential to make significant, useful discoveries improbable with the old toolset.

AI changes how we will think and imagine in future. Trillions of experiments can be run in days, for example, a fundamental gear shift in the scientific method. This kind of discovery doesn't require a prior theory or hypothesis, but arrives at conclusions through a vast process of trial and error, a kind of 'radical empiricism'. It also applies to how we invent.[28] Antibiotic resistance, for example, one of the biggest challenges in medicine, is being attacked with a similar strategy: there could be 10^{60} drug-like molecules, and finding the ones that work is a formidable challenge without AI.

AI can run countless trials, prototypes, models and design adjustments, unlocking obscure perspectives. Some years ago researchers at Aberystwyth and Cambridge Universities developed a robot called Adam, arguably the first automaton scientist capable of formulating hypotheses, running experiments and interpreting the results: studying the metabolism of yeast, and using automated lab equipment, it identified genes that coded for certain enzymes in the yeast.[29] Like his biblical equivalent, Adam is only the beginning.

AI roams far beyond science and engineering problems. Creativity was always the Promethean province of humans. But for how much longer? Machines convincingly recreate the music of J.S. Bach, the paintings of Rembrandt and the prose style of Tolkien (or BuzzFeed). AI, including GPT-3 mentioned above, produces cogent original texts, whether poetry, journalism or one-liner jokes, perfectly realised images and compelling music. Some of this simply copies humans, but some shows what is possible when the bounds of human creativity are left behind.[30]

Such work is of real aesthetic interest and value: pieces of music that last and evolve over thousands of years, or images that defy the imagination. AI isn't just about clever fakes or montages but new visual and aural languages. These are only the beginnings of a tool-engineered creative plane – a key and unprecedented difference. Indeed imagination could be one of the keys to intelligence itself, and is now a focus for AI research.[31] In the words of Yuval Noah Harari: 'If we think about art as kind of playing on the human emotional keyboard, then I think AI will very soon revolutionize art completely.'[32]

The humanities too rely on tools, showing how AI shifts understanding in unexpected places. Studying ancient history relies on a discipline called epigraphy, where scholars examine texts inscribed into durable materials like stone or ceramics: our best windows into the minds of those who lived thousands of years ago. But over the centuries these are often damaged – pieces get lost or become illegible. A field developed in which work was carried out on so-called 'restorations', hypotheses about what those missing pieces said based on a laborious study of other inscriptions. That is until researchers built a model called PYTHIA. A huge corpus of inscriptions was fed into the model, which learned to restore texts more accurately than human experts could.[33] PYTHIA and others like it help scholars go further and faster in uncovering the ancient world. While the 'digital humanities' are hardly news, they are now being supercharged by AI.

In the near term AI promises to unlock the shuttered gates of discovery; to be the fishing rod we need at this juncture in the history of ideas; to overcome the mountainous accumulation of data; to hugely accelerate the search through the space of possible ideas; to reveal unseen patterns and expand our pockets for radical thinking. We will become – are already becoming – numb to examples like these, which are just snapshots of a fast-developing general purpose tool. Whether tackling Schrödinger's equation, producing new agricultural techniques, modelling galaxies, developing investment strategies, predicting demand for wind energy or writing a symphony, ML is already a fertile producer of the next generation of big ideas, a technology capable of learning, not processing; creating, not copying.[34] But this is not the end. The notion of artificial general intelligence (AGI), a truly intelligent machine, already expands the potential for big thinking. Of necessity, we cannot be sure what such a machine would do, how or what it would think. We can however be sure that its impact would be colossal.

It might create many kinds of intelligence – each qualitatively and quantitatively different, producing new vistas on the world, capable of understanding or theorising in new modes.[35] Not just AGI then, but a cornucopia of AGIs. By definition, guessing at what these potentially diverse intelligences might address is problematic. Their potential is far grander, more autonomous and, for many, terrifying than the tools that exist to date. Theorists of AI posit that when it reaches a degree of intelligence it could start purposefully improving itself. In silicon time this evolution may play out at light speed compared to that of living organisms: recursive self-improvement will lead to an 'intelligence explosion'.[36] As the machine keeps making itself smarter on an exponential curve it will head towards superintelligence, a space of operation and understanding as far beyond human abilities as we are a microbe's.

This 'singularity' would be an event of genuine cosmic significance. Humanity would be rendered a stepping stone. In ideational terms, all bets are off. If superintelligence is ever realised, the future of big ideas is bigger than we can imagine. The universe's physical limits become the conditions of possibility. Perhaps some sufficiently advanced entity might supersede our present known limits, just as a Boeing 747 would seem to subvert all the rules of possibility available to Neolithic man (let alone to the first eukaryotic cells). The singularity is named after that of a black hole, and like it is inherently unknowable, but also intensely powerful, compelling, frightening.

Superintelligence poses existential questions of the greatest magnitude. Many doubt its plausibility. To say it is speculative is a gross understatement. It demands consideration, however. If it is achieved, human-level cognition will be a relic. A regime of intelligence which dominated the world, perhaps the universe, for 100,000 years will be over. Superintelligence would be the ultimate big idea, but maybe the last human idea. Our tools will have left us far behind. Humanity will enter the 'Novacene': a new epoch in the history of life, the age of hyperintelligence, with a new transcendent frontier altogether.[37]

Instruments of the Third Millennium

AI is only the beginning. As in the seventeenth century, a constellation of tools will alter our perception of the universe. Despite the challenges, and beyond AI, the outlines of a toolkit for the new millennium are coming into view.

We too have our telescopes. Just look at those now coming into operation: the Square Kilometer Array in South Africa, capable of generating exabytes of data; the $9.7 billion James Webb Space Telescope, with six times times the light-gathering capacity of the Hubble Space Telescope; the Extremely Large Telescope in Chile, with a 39m primary mirror gathering a hundred million times the light available to the human eye and providing images sixteen times sharper than the Hubble; Breakthrough Listen's global push for extraterrestrial intelligence; and the new generation of gravitational wave instruments among them.

Our experiments are getting bigger, yes, and costing more, but they hold out the glimmer of new worlds. Already, recent results from experiments at CERN, Brookhaven National Laboratory and Fermilab are hinting at visions of a new physics.[38] CERN now proposes the Future Circular Collider (FCC), 100km long, smashing particles together at 100 times the energy of the LHC, looping round Geneva with space to spare. Not only will it allow for a more complete understanding of the Higgs boson, but CERN physicists hope it may unlock more unexplored realms. At a projected cost of up to $27.3 billion, the key word is 'may'.

These possibilities hint at what concerted efforts could achieve. Even more important are the fundamental tools, the general purpose technologies and generative platforms – levelled-up versions of paper and printing, steam, electricity and computing. So strong is the potential, and so promising the progress, that even some early advocates of the Great Stagnation are wondering whether 'cracks' are starting to appear; that rapid progress in areas like AI and biotech is, once again, supercharging a new nexus of scientific advance and economic growth.[39]

Computers are already capable of extraordinary feats, the greatest tools of the last century. It would take a human brain 63 billion years to perform the calculations a supercomputer can crunch in a single second.[40] But Moore's Law is in trouble; sooner or later it will bump against hard physical limits.

Quantum computing opens a different front. Quantum computers use quantum phenomena like superposition – where particles exist in an indeterminate state – to massively increase computing power. Used in the right way even a small number of atoms could have more computational power than if the entire known universe was converted into a classical (e.g. non-quantum) computer.[41] At the time of writing, quantum computing is a nascent but quickly evolving art – organisations like Google and IBM are investing billions and arguing about the key threshold of 'quantum supremacy', the point when quantum computers start performing calculations impossible on classical devices. Google claims to be there already.[42] Beyond that lies the even more promising milestone of 'quantum advantage' and use in real-world applications. IBM believe this will happen within the next decade. But getting it to work at scale remains daunting – the more you scale up the more errors creep in. Error correction remains an immense challenge.[43] Occasionally quantum computing researchers worry they may be on a mission more akin to fusion than fission.

These engineering problems will be overcome. What can then be done with a quantum computer? Like AI, the motivation of scientists behind quantum computing is the making of discoveries. The big hope for quantum computing is that we will be able to simulate and fully understand complex and dynamical physical systems, in particular the intricate reactions of molecules. This would let us model systems and processes far beyond our current capacity. Quantum computing has ramifications in fields as diverse as chemistry, cryptography, materials, pharmaceuticals, financials and logistics. In truth it's almost impossible to say what results or lines of enquiry we should expect: many discoveries arise almost accidentally as a result of new tools, and that will be true of quantum computing. It is symptomatic of the new toolkit – still in process, potentially of awesome power, with results still, of necessity, perceived only dimly.

Likewise digital technology still has far to run. We've not yet had large-scale use of virtual or augmented reality (VR and AR); VR is a total immersion in digitally created worlds, the 'consensual hallucination' of William Gibson, AR an omnipresent overlay atop the physical world. These are technologies that are imagined, hyped, built, and fail time and again. It's easy to laugh at gauche forays like Google Glass, but at some stage some application, device or service will gain purchase. The sensory promise – totally immersive, intimate and fantastical – and the potential rewards are too great.

Mass adoption of VR and AR represents a revolution as large as the PC or smartphone – perhaps even writing. It's not hard to see how ideas in art, culture or policy might arise in this milieu. Ubiquitous VR might even kill the whole concept of a 'real' world. Full-scale digital universes are an inexhaustible testing ground for every conceivable form of idea. An endless taxonomy of persistent simulated worlds would be the plaything, laboratory and goldmine of the future, an infinity subject only to the constraints of the imagination. We've barely scratched the surface of what this landscape of virtual realities might be or might mean. As with all these tools, it is at once exhilarating and disturbing.

Nor is the 'real' world ignored. After all, it's in the material world of jet engines, medicines and societal resources that we have stasis. But matter itself could become a malleable platform, at first with advances in areas like additive manufacturing (3D printing) and robotics but then at a new level altogether. Nanotechnology – or atomically precise fabrication – does for the processing of matter what digital technology did for information. At an extreme, any item could be assembled quickly from the most basic building blocks.[44] Fabrication instructions are 'all' you need. Objects could thus be assembled from abundant materials with zero waste, anything unused being returned to the system's basic 'feedstock'. Our entire political economy and material culture would be transformed by the resulting abundance. It sounds like science fiction, but its proponents argue it's feasible within the laws of physics and engineering possibility. It just needs resources, and lots of them. We've already got better working at the nanoscale. And the work of something like Lee Cronin's group in Glasgow hints at the promise of new molecular and chemical machines and computers.

Perhaps the greatest significance of nanoscale engineering, arguably the most significant tool of the new millennium alongside AI, will not be materials or learning machines: it will combine both. More to the point, it will be alive.

Life is extraordinary. It's a mechanism for fighting entropy, for processing energy and information at every level from the most simple virus to planetary ecosystems; it's an atomically precise self-assembly and self-healing system. Now, for the first time in history, these features may be fully tractable.

From the beginning of the first Industrial Revolution, the development of technology with all its consequences was driven by an extraordinary convergence of physics and engineering. They bounced off one another, a textbook case of tools and ideas co-creating, practical devices unleashing new insights which in turn enabled new generations of tools, which then transformed society and the imagination. This feedback loop between physics and engineering produced everything from lightbulbs to televisions, jet engines to nuclear power, radios to smartphones: technologies underpinned by advancing comprehension of energy, electricity, the structure and behaviour of the atom.

We may be living through a similar moment, but this convergence is uniting biology and engineering.[45] And we're only at the beginning. Just as it was difficult to predict what the union of physics and engineering would produce in, say, 1760, so it is with this new synthesis.

The costs of genome sequencing (and synthesis) are falling fast, following what *The Economist* called the Carlson curve. It took $3 billion, years of work and thousands of scientific minds to sequence the first genome. Now it takes a matter of hours and costs a few hundred dollars. The falling cost of DNA analysis and synthesis, new technologies and techniques like CRISPR-Cas9 gene editing and machine learning, and, often, new medical applications, mean biological technology and knowhow will touch almost every corner of our lives. Just as digital technology provided a civilisation-wide 'platform' for new thinking, from scientific advance and creative expression to business and political possibility, rendering biology itself a platform or general purpose technology has far-reaching implications.

Gene editing is a good example. Genetic engineering dates back to 1973, when Stanley N. Cohen and Herbert Boyer (initially working on bacteria) transplanted genetic material from one organism into another. But development was slower than hoped: months could be lost on engineering embryos that never took. The 2001 mapping of the human genome was a landmark: three billion letters of genetic information decoded. Again though, it wasn't the immediate springboard many expected. Making even 'straightforward' genetic changes was still costly and time-consuming. Then came gene editing. The writer Nessa Carey explicitly likens this to the digital revolution, growing from earlier and more primitive roots: 'This new technology is cheap, incredibly easy to use, fast, flexible and may prove to the silicon chip to the Cohen and Boyer valve.'[46]

A breakthrough in 2012 made by Jennifer Doudna and Emmanuelle Charpentier, for which they won the 2020 Nobel Prize in Chemistry, ushered in an era where the genes of any organism, from bacteria to beetles, can be edited like computer code.[47] This means we can change the genomes of any living organism, including our own, with astonishing precision. It allows for fine-grained editing, but also for major changes, including alterations to the germ cells that produce eggs and sperm and from there to all 40 trillion cells of the human body: these genetic changes are permanently embedded in the germline, passed on to succeeding generations. Natural selection has been comprehensively supercharged, perhaps overcome.

Gene editing techniques like CRISPR can work on almost any species with relative ease. They thus provide a platform for curiosity, enabling imaginative experiments, catalysing knowledge and boosting scientific capability. One example of gene editing-enabled research comes from biologists studying how butterflies produce their distinctive wing patterns. The findings have gone beyond lepidoptery, shedding new light on evolutionary mechanics. But just as significant, such experiments were previously impossible. As one of the lead scientists said: 'These are experiments we could only have dreamed of years ago. The most challenging task in my career has become an undergraduate project overnight.'[48]

But it is still only the beginning – a stage akin to computers fifty years ago, with all the general-purpose promise that implies.[49] Synthetic biology is the emerging discipline of designing and building organisms. It could follow a trajectory similar to information technologies. In the 1960s semiconductor circuits were still hand-carved with a scalpel. They now exhibit a mind-boggling complexity, squeezing billions of transistors into our pockets (11 billion alone on a single iPhone CPU, performing well over a trillion operations a second). What if the construction of organisms followed a similar curve, engineered at atomic detail, moving from painstaking hand work on the lab bench to the equivalent of a smartphone-producing industrial process? Just as computer-aided design enhanced airplanes and circuits, so it will genetic engineering. We'll have tools, in other words, for programming complex function into cells, for the algorithmic design of genetic parts, including human cells. What today exists only in simulations could, thanks to nanotech and synthetic biology, happen for real – and on aggressive timelines.[50]

The envisaged applications of this bio-engineering convergence are astounding. Viruses that create batteries and proteins that purify water.[51] Cheap, personalised cancer vaccines made at scale and low cost could be another hammer to Eroom. Crops resistant to infection; microbially produced smartphone screens; livestock like pigs used for growing organ transplants, so called xenotransplantation – the pioneering Harvard scientist George Church and his lab are already overcoming hurdles here. So-called gene drives see disease-spreading insects or invasive species edited to lose reproductive capacity, whittling down their populations. Nanomedicine is coming of age, not least in the fight against Covid-19, but also in the treatment of other conditions like multiple sclerosis.

Biologicals are used at immense scale in industrial

processes, part of everything from algae-derived bioplastics to niche foodstuffs, from major cash crops to pharma, fashion (yes) and even biocomputers. That usage will increase, reordering supply chains, businesses and the consumer experience. From new sources of energy to living buildings to biological stores of information, this is an incredible tool. McKinsey estimates that 60 per cent of physical inputs to the global economy could be biologically generated, a $4 trillion business in ten years and growing from there.[52] Synthetic biology may also become instrumental in combating climate change: we might start with crops that desalinate soil, fish that survive in acidic oceans or pollution-draining plants, but surely there is a lot further to go.

The bio-frontier is likely to continue moving at pace, with massive ramifications. Again, beyond these envisaged uses lies a universe we cannot yet see or even imagine. There are a near-infinite array of possible proteins. Scientists working on biotech can no more envisage all the applications than Turing and von Neumann could have foreseen TikTok, Wikipedia and phishing attacks. But we can be confident that tools like AI or synthetic biology will not remain walled off in an airtight canister labelled 'technology'; rather they will have a major and unpredictable impact on things like markets, bureaucracies, ideologies and aesthetics.

AI and biotech may seem the opposite ends of a spectrum, but some of the most dramatic tools could arrive at their intersection, another trans-disciplinary loop with profound impact for the human mind and frontier. Over the next decades or centuries we should be able to sketch a detailed picture of how the brain's atoms, electrical activity and chemical reactions produce our minds. Existing prostheses – writing and ledgers to store information, all those telescopes and compasses building up to computers, the Internet and AI – could coalesce and transmogrify: tools from AI to virtual worlds that outsource the cognitive load; designer drugs that supercharge cognition and perception; and ultimately the fusion of organic and inorganic constructs to further evolve our intelligences and consciousness. The lines between brain and machine, artificial and biological ideas, crumble. It all sounds quite grandiose and alarming: it is.

Investment has been made in some of the world's biggest neuroscience and neurotech research programmes, like Europe's €1 billion Human Brain Project, the US's Brain Research Through Advancing Innovative Neurotechnologies (BRAIN) Initiative or the NIH's Connectome project. Academic advances dovetail with technology: neuroprosthetics will let people control wheelchairs or artificial limbs. Neurotechnology from the likes of the BrainGate research project can restore movements or communicative functions to the paralysed or those suffering from neurodegenerative diseases. Elon Musk's Neuralink and ARPA's Brain Initiative are just two leading efforts pursuing brain-to-machine interfaces: scalable, high bandwidth systems plugging brains into computers via microelectrode threads robotically sewn into the brain. This work is still at an early stage and involves opening the skull, always a delicate procedure. But others believe non-invasive brain–machine interfaces are possible within ambitious timeframes.

It doesn't require much to wonder what such technologies imply for human cognition. The very process of thinking could drastically change. Although we don't currently understand enough about genetics, intelligence or the brain, it's conceivable that we could re-engineer our brains for greater intellectual abilities even as we connect them to the cloud. Even if these technologies make you queasy, it's hard not to be intrigued about the ideas such minds might produce.

Our consciousness and thought patterns are, so far as we can tell, contingent on the evolutionary processes behind them. This must then be true (to some extent) of our ideas. Brain–machine links would be a leap above anything we have experienced before.

Fusions of the biological and the technological could also recreate the human mind *in silico* – whole brain emulations or 'ems' that reproduce a human brain neuron by neuron, all 86 billion of them complete with all 100 trillion synaptic connections – or at least achieve the same effect, hence theoretically recreating individuals as new machine-based entities. Your brain, but not *your* brain.

Such ems would be as remote from us as we are from our distant ancestors on the savannah: working on a hugely accelerated timescale (a single year may feel like millennia or more), anchored in virtual realities clustered in dense physical architectures of servers and coolants, where clans of ems reproduce themselves at will, working at jobs barely conceivable in the fleshy present.[53] These minds could run millions of times slower or faster than ours. If ems can sufficiently reprogram themselves, untold numbers could quickly evolve far from their human-like starting points to produce a bewildering ecology of minds. Em ideas, like AI ideas, are fundamentally grand and alien, even nudging towards the theological or numinous.

It's not clear where or when or whether these technologies will be deployed. Along current trajectories of scanning, modelling and programming, some think it feasible in the medium term.[54] It would represent another epochal change in the nature of consciousness, reshaping our minds, creating a cyborg species capable of, say, directly comprehending quantum phenomena, or existing at vastly different scales and speed.[55] What would it feel like to conduct trillions of calculations a second? To have, at the tip of our mind, more knowledge and processing power than exist in the world today? Like AGI, it is virtually impossible to grapple with the ideas such transhuman minds might produce, what art they might create, what social norms might arise, how their polity would be organised. Suffice to say it seems plausible that many barriers to big ideas would be irrelevant. So much for the mysterian limits on our all-too-human brains. So much, once again, for the human frontier.

Time for some cold water.

None of these tools are immune to the dynamics already described in this book. They aren't coming down the track on some automatic expressway. Instead there are likely to be unforeseen barriers, reversals, hitches, shutdowns; like the wider picture, each exhibits uneven progress. The greatest toolkit the world has ever seen is incomplete.

That work-in-progress is already delivering conceptual dividends. But it hasn't been easy. Before the boom, AI endured an unfashionable and underfunded winter. And arguably, the boom hasn't delivered. No other sector has matched the recent growth in AI investment: in just over five years, $83 billion of venture capital was poured into AI startups alone.[56] Corporate and government funding worldwide is at least as much and likely vastly more. More than 60,000 papers are published on the topic of AI every year. And yet for all that expenditure and effort we are in general still seeing a finessing of techniques first proposed well over a decade ago. Arguably the big ideas in AI preceded all this investment, which rolled them out. Even then AI often works on certain closed problems. Good in the confined space of a game, less so in the ambiguity and chaos of the real world. AI is, at present, a series of statistical inference engines, not a magic wand.

At the same time, experiments are getting more expensive and computation-intensive, doubling their hunger for computing power in just a few months. One paper suggests that if this trend continues, absent an unforeseen breakthrough, it will be sustainable for at most a decade.[57] There are growing suggestions that deep learning will start to butt against computational and technical limits (not to mention economic and environmental limits).[58] AI is getting bigger, but not making the necessary conceptual paradigm shifts; it is locked into a given developmental path even as its real-world applications are subject, in some quarters, to growing scepticism. Arguably the research field has narrowed as a diversity of approaches has given way to a blanket new orthodoxy – exactly the kind of stagnant society pressure that will ultimately slow progress.[59]

Little surprise that some in the field argue that concepts like AGI or superintelligence are red herrings, category errors, mythical constructs more than engineering goals. Like fusion, AI experts have confidently been predicting for over sixty years that AGI is, in this case, fifteen to twenty-five years away.[60] The closer you look at AI, the more it could fit the breakthrough problem and diminishing returns – exponential increases in research do not translate into commensurate progress. And of course that is to set to one side the question of to what extent this is relevant for human frontiers: if we cannot understand the concepts of our machines, are those concepts in any meaningful sense our own?

The Idea Paradox suggests we always work on harder problems. Only an escalation of our tools ensures we keep up. Yet the tools themselves are, in many instances, massively complex, caught in the same dynamic of escalating difficulty. Nor is society some automatic springboard. Instead we create these tools freighted with all the problems and pressures outlined in the last chapter. They are expensive, risky and slow to develop.

It won't escape the reader's notice that the twenty-first century's most significant tools present acute social, political, economic and ethical dilemmas. Aside from fears over automation, AI has problems of safety and control, explainability, value alignment and unintended consequences. Failure to address these issues has serious implications. Genetic engineering makes people uneasy: they ask whether new capabilities will be programmed down the germline, echoing through generations, entrenching an artificial inequality in DNA itself. Whether the richest will get everything, while different countries follow different regulatory paths. Above it all, and in the wake of Covid-19, the prospect of bioterrorism fills me with an unshakeable dread. And yet without an advance of these tools, we won't have an answer to the next pandemic. There are no simple paths.

Unlike telescopes or the personal computer, tools and techniques like AI and synthetic biology come parcelled with a degree of existential risk. That should give us – at the very least – pause for thought. Some unsurprisingly argue we should stop working on these rightly contested technologies altogether (more on that in the final chapter).

Society could intervene in other ways. Gene editing is a good example. While the science of CRISPR proceeds apace and companies develop new medical treatments, the field has been dogged by uneven regulation and battles over IP. As knowledge galloped ahead, Berkeley (where Jennifer Doudna is based) and the Massachusetts-based Broad Institute slugged it out in the courts. Gene editing is ready to change the world, but a combination of over-regulation in some territories, patent wars and scientific nationalism weighs down on the field. Meanwhile VR has been around for years, but to date no one has found a fit between the product and its market. Perhaps it stands poised to launch a bright multiverse of possibility; perhaps most of us are content with TV. Major tools are subject to a vicious, recurring hype cycle.

And does the financialised and post-Covid world have the stomach for breakthroughs' ever-increasing cost? To date the US government has spent billions on the National Ignition Facility, a laser research institute, at Lawrence Livermore National Laboratory with, in the words of one commentator, 'no tangible results in sight'.[61] The public and government appetite for such expenditures, epitomised by CERN's new $27 billion particle accelerator, is shaky. A major economic downturn makes government and business investment in high-risk areas of research even more tenuous. It may be down to the tech giants, almost singlehandedly responsible as they are for much of the investment in these tools, to maintain current levels of R&D – if they even exist to do so.

Build this new toolkit and we can realise the next generation of big ideas, and smash the limits of possibility. But threading the needle to successfully, safely complete the toolkit is no done deal. As with AI, the whole setup is still incomplete and uncertain. This explains our mixed record; why we don't have colonies on Mars, flying cars or fully fledged virtual worlds. We're still on the journey, not at the destination.

Our place in the history of ideas helps keep this in mind. In the deep background are the Cognitive Revolution (which occurred around 70,000 years ago, introducing abstract thinking, language, art and imagination and complex social norms), the Agrarian Revolution (10–12,000 years ago) and a Textual Revolution (when writing was developed, 5000 years or so ago). The modern era can then be divided into three. There was the period of the Renaissance, the Scientific Revolution, the Enlightenment – an era lasting roughly three hundred years from 1470 to 1770. This was followed by a Long Industrial Revolution. Zooming out, the two centuries from 1770 to 1970 become one continuous story of the world's leading edge moving from pre-industrial to fully industrialised. Within that the world had smaller revolutions – the 1IR and 2IR and their subsets: from water mills to steam factories to Fordist oil- or electricity-powered factories; from canals to trains to cars to planes. But they are part of a broader pattern.

Around 1970 something new began with de-industrialising, de-materialising economies and a slowing growth after the end of the 'special century'.[62] A shift became evident in technology, capitalism, society and culture.[63] ARPANET, the precursor to the Internet, went live in 1969. Intel's first microprocessor was launched in November 1971, two hundred years after Richard Arkwright's mill at Cromford in Derbyshire ignited the Industrial Revolution.[64] Two years later came the first instance of genetic engineering. A year after that, Vint Cerf and Bob Kahn published the TCP/IP protocols governing the Internet. The Information Age was born.

In this context it's premature to talk about a 3IR and especially a 4IR, overly narrow periodisations of the present and future.[65] Rather, both are instances of a new era that began around 1970 and is still in progress. We are in the opening phase of the era that will replace the Industrial Age. In this schema that was a two-hundred-year event, a hundred years less than the Scientific Revolution. This post-industrial, digital, information-rich era has been around for fifty or so years; I think of that as at most halfway through its probable life.

This revolution cannot yet be judged *in toto* and stagnation has to be seen in that context.[66] As with human rights in the early twentieth century, there will be setbacks, but also plenty more to come. Forty years after the invention of the printing press, books still resembled scribal manuscripts. Only when printers like the Venetian Aldus Manutius learned to redesign the codex, its fonts and layout and sizing, enabling large-scale print runs, did it become the great catalyst of the Renaissance, just as it took decades for manufacturers to reorientate around steam and then electricity. The 3IR or 4IR aren't historical artefacts; developments in AI tell us they are dynamic events still being written. Look back at the technology: it started with computing, with big mainframes that clunked through data, before hitting an inflection point after coupling with a communication medium to form the Internet.[67] Now we're adding a third component, learning. In future we might add another – real intelligence, whatever that may turn out to be.

The opening decades of the twenty-first century are caught in the middle of a grand unfolding. The world's new toolkit – and the revolution it both epitomises and enables – is still being assembled, and will be for decades more. Already we can begin to see how these tools might help produce the defining ideas of our time, ascend the next rungs of the Idea Paradox, definitively buck any sense of stagnation, find the links, combinations and recombinations we cannot see unaided. The real question is not one of form or potential; it's whether we can build them at all, and if so make them safe and sustainable.

That in turn relies on the kinds of societies, cultures, institutions and organisations that give rise to new ideas. Is there another side to civilisation today, one that bucks the stagnant society? What does it look like? Is it capable of building the toolkit? Those questions are explored in the next chapter. Again, there are promising signs.

8

The Great Convergence

Scaling Up

The scientific world was stunned. A hastily assembled press conference in Hong Kong on 28 November 2018 announced the birth of twin girls in Guangdong province, China. They were (to the public at least) called Lulu and Nana.

It should have been a happy event. But something by turns remarkable and appalling had come to light, leaked a few days earlier in an exposé by the *MIT Technology Review.* Using the 'molecular scissors' of CRISPR, those innocent twins were the first humans born with genetic material directly manipulated by scientists. He Jiankui of Shenzhen's Southern University of Science and Technology had taken the most advanced gene-editing techniques and applied them to the DNA of Lulu and Nana – to human life.

Planning began in June 2016, and intensified in 2017 with the recruitment of couples.[1] In early 2018 the girls had been born. When they were just small bundles of cells He had, in vitro, edited their genomes. A threshold had been breached.

Professor He, who had spent years in the US before

returning to China, initially touted his work's putative noble purpose, editing a gene, CCR5, that was connected to HIV. He was 'proud' of the results. The girls were born 'happy and healthy'. Another pregnancy was already on the way and the project had more in the pipeline. On the face of it, here was a major twenty-first century breakthrough.

The outcry was immediate. With CRISPR creators Doudna and Charpentier in the vanguard, the scientific community were unconditional in their condemnation. Doudna was 'horrified and stunned'. What ethical safeguards or considerations had been in place? Where was the legal framework, the international collaboration on a study of the utmost consequence? Safety tests had been inadequate, procedures for finding patients ignored, ethical review documents forged. It soon emerged that both the university and the Chinese authorities claimed not to know about He's work. They quickly disowned him, first putting him on leave before a criminal trial resulted in prison and a fine of 3 million yuan.

The process and results had not been peer-reviewed.[2] Problems were apparent in the science. There are more pressing reasons to edit a human genome. He Jiankui also appeared to work sloppily, perhaps not even helping to impart immunity to HIV as was intended. Not only was the work inessential, but it could make the girls more susceptible to other conditions. One ethicist described the experiment as 'monstrous'. This wasn't the breakthrough people had been hoping for; it was a nightmare from a rogue scientist playing God.

Media coverage focused on the moral and biotechnological implications. It crystallised fears about the ethics of the new toolkit and highlighted problems with an uncomplicated charge for big breakthroughs. History had been made; but it wasn't, said the leading scientists, meant to happen like this.

There is another significant aspect to the He Jiankui affair:

its location. In the twentieth and early twenty-first centuries breakthroughs were common in the kind of places that gave rise to CRISPR: places like Berkeley, California or Cambridge, Massachusetts. Shenzhen, where He conducted the trials, didn't even exist until 1979; but when Lulu and Nana were born it was a megacity of 12.5 million people, host to the Chinese tech industry and now the first gene-edited human beings. It hints at something much wider than one experiment. The geography of ideas is shifting more rapidly than ever before; a new phase in the history of ideas is about to overturn our extant models and assumptions.

For those following biotech, the existence of the twins and their place of origin were no surprise. Although researchers in the US and UK soon published similar work, a team from Guangzhou had, in March 2015, been the first to edit the genes of a human embryo.

For several decades, China had been positioning itself to lead the field. In the 1990s and early 2000s China contributed about 1 per cent to the Human Genome Project through its Beijing Genomic Institute. But in subsequent years BGI had become the world's biggest genomics research centre, its genome sequencing capacity equivalent to that of the entire United States. It occupies an ambiguous but advantageous space between public and private – supported by a multi-billion-dollar credit line from the Chinese government, but also capable of raising significant capital from the private sector.[3]

Western scientists had long viewed Beijing's attitude to scientific norms and ethical safeguards as lax. Compared to European or US regulators, the Chinese green-light projects faster and with less scrutiny. Despite the government's disavowal of He, China's regulation of gene editing was labelled 'permissive' by *Science*. China takes a less risk-averse view than even the US, creating what appears to be a lighter-touch regulatory environment. No one in biological circles is quite sure where Chinese technology has got to; unlike in most Western contexts, there is a sense that much important work, including methods and clinical results, is never published. There are even suggestions more than a hundred people have undergone gene editing.[4]

The Chinese government knows the US derives immense power, prestige and wealth from being the epicentre of digital technology. If biotech is another world-changing general purpose technology, they want in. Overall the lesson of history is clear. The West shouldn't have a monopoly in setting the pace of discovery and research; the world's new toolkit should – must – be built in the East. A new reality emerges – a genuinely multipolar R&D environment.

The frontier has, for most of history, for most people, been closed. If the eighteenth century marked a Great Divergence, when Europe pulled away, we are now living in the snapback: the Great Convergence. Ready or not, we are seeing a convergence of humanity at the forefront of thought.[5] This is where we will build our new tools. Its consequences will shape everything about ideas for decades to come.

The rise of China and the levelling of nations is usually presented as an economic or geopolitical fact. But this misses its full significance. It is also a pan-civilisational change in our capacity to create, nurture and give purchase to ideas. This shift lags behind economic development by years. Only now and in ensuing decades will we experience the Great Convergence's full impact.

For much of history China led the world. A cradle of agricultural civilisation, it sees within itself an unbroken cultural thread dating back thousands of years. For much of that time it had the world's largest economy. Even in the eighteenth century, as the 1IR started to transform Britain, China had a larger economy than the major European powers. Most people in China, and indeed Asia, do not find it surprising to see their homelands thrown back to the forefront.

For two centuries the West overlooked China's historic role as a crucible of humanity's biggest ideas. This error becomes more obvious by the day. Under the Zhou dynasty of the first millennium BCE China made the transition from bronze to iron, pioneered hydrology and mathematics (including the decimal system), and made silk. The Warring States period gave us sages like Sun Tzu and the advent of Daoism. The Han dynasty (202 BCE–220 CE) cemented the Confucian ethic.

That was all before the 'Four Great Inventions': gunpowder, the compass, paper and printing, the glories of ancient China. Francis Bacon held them up as the world's most significant inventions, not an unusual thought either at the time or since. Revisionist historians have questioned the Four, but mainly on the grounds there are so many other significant Chinese inventions it hardly makes sense to focus on only so small a number.

Just as Western development hit a post-Roman low, Asia scaled new peaks epitomised by the golden age of the Islamic caliphate – the Abbasid dynasty in Baghdad created a House of Wisdom, advanced astronomy, medicine and mathematics, used paper and developed the classics of Greece and Rome, just as the Tang (618–907 CE) and Song (960–1279 CE) dynasties saw a new flowering of culture and learning in China. While Europe floundered, China built the Grand Canal; it accounted for a quarter of global population, living at much greater densities thanks to its superior agricultural techniques and technologies. It led the world in crafts and manufacturing, enjoyed trans-continental mercantile links and supported a class of scholar-bureaucrats who spent their time governing and thinking. By the tenth century the Tang's imperial library consisted of some eighty thousand volumes; Europe's largest contemporary library, in a Swiss monastery, had eight hundred.[6] They built vast bridges, buildings and ocean-going junks, wove sumptuous cloth and manufactured fine porcelain, assembled sophisticated clockwork mechanisms and irrigation systems, and disseminated knowledge through printing and paper – in all, Tang and Song China and its Islamic peers exhibited organisation and pushed innovation wholly beyond contemporary Europeans.

But by the late nineteenth century circumstances had changed. Qing China was overpopulated and impoverished, lacking farm machinery or fertiliser to improve productivity and feed its people. Its governance was riven with distrust. The Taiping rebellion of 1850–64 devastated swathes of the country and was only one revolt of many. Outside a few isolated examples, it lacked the accoutrements of modernity like railways, telegraphs, factories, mines and ironclad ships.[7]

The archaeologist and historian Ian Morris puts this into perspective.[8] Using an index of social development, he benchmarks civilisations through data like size of their largest city, access to energy and means of information storage. He focuses on what he regards as the two cores of civilisation: one starting in around 9000 BCE in the Fertile Crescent of south-west Asia and migrating westwards over time, the other forming two thousand years later between the Yellow and Yangtze rivers. At times the civilisations were close – for example in the overlapping period of the Roman and Han empires – but there was never a period when both simultaneously peaked while in regular communication: that is, there was never a global push on the frontier. While the Western core held the advantage for much of early history, Morris shows that between 550 CE and around 1750 the Eastern core dominated even as that fertile convergence remained elusive. Until now.

Throughout history societies have been 'prisoners of geography', constrained by place, culture, knowledge and resources. Now, rusting link by rusting link, the shackles are breaking down. Morris's index shows development rising 900 points from Neolithic man to the age of nuclear power and the computer. But if trends continue it is forecast to rise an astonishing 4000 points over the twenty-first century. In the next decades we may see more development than in thousands of years. 'What a mockery, this, of history!' he writes.[9] Deep historical patterns are retiring. Old categories like developed and developing economies, assumptions about the scope of progress, the deep traditional localisation of innovation, are at the very frontier rendered irrelevant. China is the most obvious example of how, over the next century, big ideas will disperse around the world, no longer concentrated but rather drawing on the deep well of humanity everywhere.

Several events have been key to this Sino-oriented transition. The Green Revolution fed the world and, after the Great Famine, Chinese peasants among them. Deng Xiaoping's reforms (which were in truth often bottom up) turned the axis of history. The Berlin Wall fell and the world was brought into one economic system just as the Internet stitched it together in a single information-rich communications superstructure.

The results are readily apparent. In 1990, 60 per cent of global trade was between rich countries, while that between developing countries was just 6 per cent. But since then, the latter has grown twice as fast.[10] Asia took the lion's share of global growth over the last forty years. By 2012 China was the world's largest manufacturer. In the twenty-five years after 1990, the dollar value of its exports grew forty-fold and it continues to rise steeply.[11] Its high-tech businesses, like Tencent, Bytedance, Alibaba, Didi and DJI, eclipse all but a few rivals in market cap, R&D spend, user numbers and the deployment of groundbreaking tech. From the 2010s on, no one could claim new technologies were the privileged domain of the West.

Chinese research is well-resourced and increasingly world-leading, not least in building the new toolkit. China spends $500 billion or more per year on R&D, significantly more than the EU and closing on the US (which it will have likely overtaken by the time you read this).[12] Between 2000 and 2016 China's share of global scientific output quadrupled; since 2000 it alone has accounted for a third of global growth in R&D and has seen the second greatest increase in R&D intensity after South Korea. 2016 was a litmus year: for the first time China published more scientific papers than the USA to become the world's leading producer of scientific knowledge (in quantity – but quality, as measured by citations, also improved four-fold).[13] A study from Elsevier and Nikkei found that in twenty-three out of thirty 'hot' fields, Chinese researchers published the most highly cited papers.[14] It has a larger share of global patents and of the global STEM workforce than the former leader, the US; indeed almost 50 per cent of global patent families are going to Chinese inventors.[15] That work-force is better educated than ever: between 1990 and 2010 the number of college graduates grew tenfold to 8 million per year, while the number of Chinese postgraduate degree holders grew fifteen-fold in the same period, again surpassing the US totals, even as Chinese universities shot up the world rankings.[16]

This boils down to specific advances beyond biotech. Andrew Ng, the AI pioneer, argues that the complexity of the Chinese language, and its level of investment, have pushed AI natural language processing ahead of the West, while Eric Schmidt, former CEO of Google, expects China to overtake the US in AI in the near future.[17] Whereas the UK is putting at most £150 million towards the development of quantum computing, China has invested $15 billion and rising – its National Laboratory for Quantum Information Sciences is the largest anywhere in the world. Projects like the Chang'e 4 probe, the world's first to land and explore on the far side of the moon, the 500m Aperture Spherical Radio Telescope, and advances in cleantech like batteries and photovoltaics are the fruits of year-on-year spending increases of well over 12 per cent. A single research park in Beijing, Zhongguancun, sprawls over 100 square kilometres and hosts 20,000 companies. The Chinese news agency Xinhua reports that a base is planned for the Moon's south pole, and a mission to land on Jupiter's moon Callisto is also in the works.[18] It's hard to see any of this grinding to a halt in the next decades. There is nothing like this trajectory in the West and, if it continues, it is likely a matter of time before China becomes the leading nation in science and technology.

China has also pursued a new brand of state-commercial digitally enabled one-party governance: a distinctive twenty-first century twist on old governance models and with it the blueprint of a challenging new social and political order. It has become a testbed for architectural and engineering extremity and experimentation. Its tech and cultural businesses have pioneered new forms of digital culture and participation. Its long-term Thousand Talents programme sucks in star researchers and performers. Five hundred Confucius Institutes have been built around the world and its response to the Covid-19 pandemic left Western nations fuming – but flailing. The Belt and Road Initiative is a unique multi-trillion-dollar globe-spanning infrastructure project of unprecedented scale – a *bona fide* twenty-first century big idea. Note that this picture is not just about the injection of economic resources: it is about ambition, a set of values and a broad culture hungry to deliver new ideas. Such an environment is no longer a Western anomaly.

This is what a multipolar world looks like. We need to understand what it means, not just for political power plays or economic growth but for the very stuff of ideas that underwrites it all. Innocent Lulu and Nana will not be outlier breakthroughs, even if they are hopefully rare examples of unwarranted human experimentation.

No one could doubt Asia's centrality to the future, with its nations of vast land mass and the world's most populous countries – not just China and India but demographic giants including Indonesia, Pakistan and the Philippines. States like Singapore or Qatar have among the highest per capita income anywhere on Earth. The continent has growing institutional clout – see the success of ASEAN or the upswell around the Asian Infrastructure Investment Bank. Asian states like China, Singapore, South Korea, Hong Kong, Taiwan and Japan dominate the PISA educational rankings, robot installations and high-tech industries like semiconductor fabrication. With every passing decade of the twenty-first century its potential is further realised.

Aside from East Asia, an arc from Pakistan and Iran to Indonesia and India encompasses young, dynamic populations of some 2.5 billion, poised to hit the front lines of the economy.[19] Collectively there is scope not just to catch up with, but to leapfrog the West. Either way, it portends the next phase in the history of ideas.

India is well placed to follow China – it will be more capable than ever before, more influential, with resources equivalent to its European, American or East Asian peers operating on an immense human scale. India, like China, has a partially obscured legacy in Western eyes (it all too often stops at yoga, Bollywood and curry). Under the Gupta empire of the turn of third and fourth centuries CE, it was an intellectual giant – completing the *Mahabharata*, inventing the concept of zero and chess, exploring Buddhism at the great Nalanda

University. For five centuries until the eighteenth, it was the world's great manufacturing hub. Like China, it began faring less well as the European nations rose.

In just a couple of decades, India has built a $100 billion IT export industry.[20] It has a rapidly growing set of unicorn tech businesses like Paytm, and has the capacity to close the gap with China and the US. Although they are yet to match the world's best universities, the elite Indian Institutes of Technology churn out an eye-opening proportion of the world's top engineers, programmers and mathematicians, just as the country still produces a large proportion of the world's top doctors. The CEOs of world-leading companies like Microsoft, Alphabet, Adobe and IBM are Indian. India implemented technology at scale like Aadhaar, the world's largest biometric identity card system built on iris recognition software, and is now looking at a suite of digital services, IndiaStack, built on top of it.

What's more, India still reaps a demographic dividend no longer available to an ageing China. In the years to 2025, 170 million will join the Indian labour force.[21] In the 2010s India embarked on a major process of renewal, focusing on infrastructure and regulation. It wants, like China, to grow not just first-tier cities, but to establish 'second-tier' cities on the leading edge; Pune, Hyderabad, and the girdle of towns around Delhi are already there. Meanwhile India has pioneered *jugaad* or frugal innovation, which is about innovating with less for people who can afford less; itself a big idea about how we can pursue progress. Companies like GE and Siemens have set up R&D hubs in India with a view to building a new generation of lower-cost medical technologies – *jugaad* in action.[22] Its Jio low-cost mobile network put Internet connectivity in the hands of hundreds of millions overnight. The Indian Space Research Organisation is planning a mission to Venus, Shukrayaan-1, which could be instrumental in confirming the hypothesis that life is present in the planet's atmosphere.

Young, comparatively open and ambitious, India couples this with long traditions of enterprise, medicine and academic achievement. With its distinctive norms, culture and governance, India will present intriguing contrasts with the US or China. The potential is clear and tantalising. At present India's R&D is small – $67 billion compared to Korea's $120 billion or Japan's $190 billion, with China above them all. But that will change fast. We might then see something like the vision from a novel like *River of Gods*: bleeding-edge tech and social formations with distinctively Indian twists; AI and AR software manifested as Hindu gods, biotech that takes a lead from ancient Indian medicine and cultural practice, different values and morphologies for everything from robotics to regulation.

But this is about more than significant giants like India or China, as if a new United States or EU entered our cultural and scientific imaginations, their populations finally able explore their potential. It is a global process.

Japan has of course long been on the front lines. It remains so, with excellent capabilities in robotics, clean energy, biotech, the internet of things, automation and care technologies. It is home to the world's largest VC fund, and has a forward-looking population of tech users. Southeast Asia is on the rise – growing fast, home to almost 700 million people, knitted together via ASEAN, the next manufacturing powerhouse. The glitzy petromonarchies of the Gulf bring their own brand to the equation.

Latin America bursts with promise – to date many of its universities and companies have not had the global impact one might expect from a large, sophisticated, vibrantly creative megaregion. But the signs of change are apparent: Brazil has for example seen a huge spike in patent applications over the last decade. Many visits over the years have taught me it would be a mistake to discount the impact of Latin America during the twenty-first century. Perhaps more than anywhere else I think it will surprise us; the continent that gave us everything from magical realism to liberation theology will play a massive role. It's genuinely exciting to think of what new ideas will emerge from this buzzing, unique region.

Africa too is changing. Since 2000 six of the top ten fastest-growing economies have been African. The proportion classed as extremely poor has halved since the 1990s.[23] Growth is forecast to continue and the demographic boost is even larger than in India: 600 million will be added to the labour force by 2040.[24] Nigeria, Africa's most populous country, will be a crucial part of this. Real income per head has almost doubled in the past twenty years while the proportion who are very poor has fallen from 60 per cent to 30 per cent of the population.[25] Meanwhile that population has boomed, roughly doubling in the period 1990–2020. But that is nothing compared to what comes next – half a billion Nigerians will be added over the twenty-first century, more than in any other country.[26] Lagos has become the great finance, media and tech hub of sub-Saharan Africa, a continental-scale hotbed of entrepreneurial ambition. The challenges are daunting: a one-dimensional economy, energy shortages (ironic, given its oil), endemic poverty. All of this will require serious investment in institutions and infrastructure if it is to come good. Nonetheless, it is an exciting prospect: countries like Nigeria taking their place as knowledge-producing cultural superpowers.

There are important political, commercial and cultural aspects to this process. Asia tends towards more technocratic governance than the West and we could see a new parallel evolution of governmental forms. China might find a new form in a sort of technologically super-charged neo-Confucianism.

Ultra-successful city states such as Singapore might hint at a smaller, nimbler, more effective polity than the lumbering nation states of the past two centuries. If there is to be a postliberal, perhaps even a postcapitalist order it will likely, for good or ill, be forged in Asia. However things develop, the decisive new ideas are more likely to emerge in this arena than the ideological vacuum of the US or Europe, whose tired constitutional and social infrastructures are lurching towards breaking point. It is noteworthy that China's permissive regulatory regime in science and technology is already making an impact even where it is officially disowned.

Outside the usual Western channels a colourful trade has transformed the cultural landscape. Precision-tooled Korean K-pop; Bollywood musicals and Nollywood films; Japanese anime; Brazilian and Mexican telenovelas; Turkish dizi, popular and exportable historical epics; competitive gaming competitions dominated by Ukrainian teens; Chinese social media smashes like TikTok or WeChat. Now that the West no longer dominates culture, the creative menu is exploding – readers around the world enjoy Haruki Murakami, Han Kang, Mario Vargas Llosa and Orhan Pamuk. Filmmakers like Bong Joon Ho, Alejandro González Iñárritu or Alfonso Cuarón, video game visionaries and auteurs like Hideo Kojima and Shigeru Miyamoto, Nobel-winning life scientists like Shinya Yamanaka and Tu Youyou (China's first woman to win such an award, its first for physiology).

Nothing changes overnight. To date American universities have still won more Nobel Prizes than the next twenty-nine countries put together.[27] But how long will the ratio hold? That productivity won't suddenly judder to a halt, but we should expect the balance to start shifting in the coming decades. It has shifted before. In the early twentieth century US science trailed behind Europe. Prior to 1930, the US won only

6 per cent of the scientific Nobel Prizes. That didn't last – in the 1990s, US winners accounted for 72 per cent of the total that decade.[28] US scientists today are no different to French, British and German scientists of the 1920s.

We've seen how wide-ranging and pervasive are the forces of stagnation. Facing off against them are countervailing pressures of expansion, a colossal redistribution of opportunity, channelled in an historically unparalleled degree towards the frontier. This tension will decide what happens to our new tools – and, by extension, how high we climb the ladder of possible ideas.

Scaling up isn't just about countries converging at the frontier. Other forms of marginalisation have been just as harmful, the position of women high among them. For most of history half of humanity was effectively shut off from the frontier. Even if you set aside the immense moral import and judge it by efficiency, it's clear that for self-inflicted reasons, humankind operated with one hand tied behind its back.

This situation began changing decades ago, albeit unevenly, and although no one could remotely say the problem has lifted, almost everywhere the direction of travel is more promising than not. Education levels have been generally increasing, but much faster for girls.[29] Worldwide, the majority of graduates are now women.[30] Fertility rates have fallen rapidly almost everywhere, converging towards two children per couple, freeing women from a cycle of pregnancy and child rearing.[31] Previously locked doors, forbidden paths and glass ceilings have been unlocked, forged and smashed.

The picture is not perfect, of course. Fields from economics to coding are still widely felt to be exclusionary to women. While the position of women has improved, even in developed economies only a third of researchers are women.[32] Just under 6 per cent of Nobel laureates are women at the time of writing – about the same number that are associated with Cornell or Humboldt universities. Women of colour are even more under-represented. Women still only constitute around a quarter of professors at British universities, a historic high but still far from representative of the population.[33] But there are positive signs: 49 per cent of tenure track positions in the US, Canada and Australia are held by women, auguring a more equal future.[34]

Still only 20 per cent of the companies raising capital have a female founder.[35] The change, again, though, is going in the right direction, with the percentage doubling since 2009. The number of startups founded by women quadrupled over the same period. In just one year, 2019, Crunchbase reported that twenty-one new 'unicorns' had been founded by women. Before 2018 this figure had always been in single digits. There is still a long way to go, but reasons to feel positive. As with the wider global picture, we should see more potential being unlocked, more perspectives brought to bear; more people able, finally, to contribute to ideas.

Another feature of the convergence is the continuing growth of cities. Since their earliest days cities have been the creative locus of our most important ideas, home to accumulations of knowledge and capital, auto-catalytic centrifuges for serendipitous linkages, agglomerations of trades and cross-fertilisations. Despite some signs that remote working may dent their importance, they remain the world's recombinant factories, its engines of idea diffusion. Clusters within cities (think Silicon Valley or the City of London or Hollywood) are the key sites for knowledge spillovers, a central plank of the theory that ideas underwrite economic growth: inventors are, for example, more likely to cite patents from inventors in the same city.[36] Without urban clustering, the US would produce around 11 per cent fewer patents per year.[37]

In 1800 only 2 per cent of the population was urbanised. It was just 5 per cent in 1900 and 30 per cent in 1950. The majority of humans have been urban only since 2008, not quite long enough to feel the full impact.[38] Nonetheless, formidable change is being wrought around us: according to the United Nations, over one hundred new cities will have more than a million people and ten will grow to more than 10 million in the next decade alone, including Ahmedabad, Hyderabad, Bogota and Johannesburg, the fastest changes yet.[39] A process of urbanisation that took two centuries in Europe is happening at warp speed. Moreover many promising cities are not the familiar brand-name boomtowns – Shanghai and São Paulo, or even Lagos. Rather, by the century's end, cities like Dhaka, Khartoum, Kinshasa, Dar Es Salaam, Kabul, Baghdad and Karachi will be the largest global megacities, full of challenge and opportunity. Architectural and urban experimentation at scale is much more likely to be found on the Arabian peninsula and in the Indonesian archipelago than the old world.[40] If we're only now feeling the full impact of the last forty years, it will be some time yet before we feel all this.

A cursory reading of history suggests that significant metropolises from New York to New Delhi play an outsized role in delivering big ideas in particular, but there is more than just supposition to support the notion that they will deliver big ideas.[41] Cities exhibit 'superlinear' scaling, which makes them supercreative. Double the size of a city and you more than double the overall wealth, income, number of patents, number of universities or creative people.

Those clusters are also more amenable to radical and different ideas at the frontier. Analysis of the patent record suggests that the larger the city, the more likely are patents generated in that city to rely on newer inventions; that is, work in the biggest cities is closer to the frontier than work elsewhere.[42]

And while a separate patent analysis shows that there is a surprisingly even spread in patenting across cities of different sizes, the most radical, unconventional patents come from the biggest cities.[43] These are inventions that rely on bridging distant and cutting-edge ideas. As home to higher proportions of young and diverse firms and organisations, with more new occupations and college graduates, this happens more easily in the great flows of the city.[44]

This global picture is also about connectivity. The twin forces of economic growth and web-enabled connectivity are the deep architects of the changes described in this chapter. Drone, satellite and even balloon technology will keep the roll-out of connectivity going. By end of the 2030s, 90 per cent of people will be connected to the Internet through phones and other 'smart' devices like watches.[45] It is an underdiscussed fact that billions of people are only in recent years and over coming decades joining the Internet, and that this transforms the global distribution of knowledge and opportunity. In the words of entrepreneur Patrick Collison, 'Several billion people recently immigrated to the world's most vibrant city and the system hasn't yet equilibrated.'[46]

The scaled-up world accommodates scaled-up flows. Of information, clearly, but also people: until the pandemic, international travel grew 600 per cent over recent decades. Migration helps brave people achieve their potential. Over the century, it will probably continue to increase, even if coronavirus has delivered a seismic blow in the short term. Immigrants and their children are responsible for many great businesses of our time: Google, Apple, Amazon, Intel, PayPal, Tesla, 3M and half of unicorn startups.[47] Over one-third of US postgrads in STEM subjects are from abroad, more at the best schools.[48] Even small increases in numbers of immigrant graduates are associated with pronounced boosts to innovation as measured by patent output.[49] A world of mobile talent is one better primed for big thinking, risk taking and productive disruption.

Here is a world scaling up. An accelerated generation of big ideas is an externality of this process. As we entered the new era after 1970, nations and peoples began to converge on the frontier. Opportunities for the previously marginalised started appearing in burgeoning cities, connected digital spaces, opening economies and recently empowered universities. But only in the second and third decades of the twenty-first century has this fully begun to manifest itself in terms of the frontier. Ideas take time. Despite decades of growth and development, it was only after the Second World War that the US hit full pre-eminence. So too with the rising intellectual powers of Asia and elsewhere.

From the earliest times work at the frontier was always uneven, fragile, highly localised; tiny speartips of a hungry humanity without much time or energy to spare. Unevenness and fragility haven't been left behind – but humanity has come much, much further than ever before. How does it change the prospects for big ideas?

A World at Work

In Part I we asked why, given the scale of available resources, humanity hasn't gone further. Perhaps the answer is that those resources were simply insufficient.

Instead, the blunt mass of scaling up will make a difference.[50] Since the dawn of classical economics, size has been understood as a critical ingredient in growth – the bigger the economy, the greater the degree of possible specialisation and division of labour, the greater the productivity growth and, thus, economic growth. There is now a comparable understanding of the relationship between population and the nature and complexity of a society's technology. Throughout history, population size and interconnectedness have spurred technology. The Nobel laureate Michael Kremer famously tracked this relationship back to the year 1 million BC and found a strong correlation between population growth and technological change.[51]

More recent research supports the idea that large populations favour invention and innovation – the more people there are, the greater the chance inventions will be made, as there is simply more chance for recombinations and lucky errors to occur, and for them to reach others.[52] But the level of cultural interconnectedness and sociality is crucial to the evolution and transmission of new ideas – the more interconnected a society, the more those inventions can spread. We learn ideas through imitation, by copying what seems to work – and the more interconnected we are, the more opportunity there is for encountering things apparently worth copying.

Diffusion, gaining purchase, means lots and lots of people interacting and adopting ideas and behaviours, or the technologies based on such activity. Transportation and communication technology should both spur further invention and innovation, by stimulating more and better person-to-person contacts. Horses and writing were good in this regard; but aeroplanes and the Internet should be much better. Historically, places without connections and with small populations, like Tasmania, experience reverse evolution, where tools and knowledge are lost. Pacific islands with larger populations have many more tools than their smaller counterparts.

Given that on a technological plane we form an integrated, global society composed of, at the time of writing, around eight billion individuals, this fact alone surely is counting and will only count more. It can be easily forgotten quite what a demographic impact China alone could have on ideas. In terms of population, China is bigger than the US and Canada, the entirety of Europe, the rest of East Asia (Japan, South Korea, Taiwan) and Australia combined.

To get a more granular sense of the change, look at software developers, a central category of knowledge worker today. Their numbers are forecast to almost double over the next decade, from around 23 million to 45 million; East Asia will experience the quickest growth, followed by Latin America, while India will replace the US as the country with the most.[53] Forty-plus years of human capital development will effectively be compressed into the next ten. A similar trend holds for almost any idea-related profession. Likewise there is a considerable body of research that relates the size of an individual's or organisation's network to its ability to generate new and valuable ideas.[54] Bigger cities, populations and more connectivity mean larger social networks.

Historically, those responsible for most big ideas were a tiny elite. Even today this remains true, with some exceptions. But the pool has immensely expanded. The funnel down to that small band has been widened to (almost) encompass the globe; the winnowing effects of reaching the frontier will still leave more and more room for more painters and quantum theorists, inventors and entrepreneurs.[55] Consider that on the cusp of the French Revolution, the Republic of Letters at its maximal extent formed an educated epistolary network of scholars, poets, antiquaries, librarians, natural philosophers, bureaucrats, courtiers, diplomats, lawyers and enlightened clergy of some thirty thousand.[56] Already that reflected a phenomenal increase on the sixteenth century, when it numbered only around 1200.[57] But it would not be unreasonable to suppose that today's global network might encompass some 30 million individuals, just a fraction of the world's population, but a thousand-fold increase on the late eighteenth century, and vastly more again on earlier times.[58]

A study of exceptionally talented students in the International Mathematics Olympiad showed first how important these individuals were for coming up with new ideas: achievements in the field of maths were correlated with achievements at the Olympiad. But second, how hailing from poorer countries held these students back.[59] Coming from a poor or middle-income nation was correlated, thanks to diminished education, infrastructure and opportunity, with lower lifetime research output and impact. Such students, ultra-gifted though they might be, were simply less likely to become knowledge producers. Had the study group all hailed from rich countries, their research output as a whole would have been 17 per cent higher. In short, for the generation of big ideas, talent matters and geography matters. As countries scale up, the impact of more geniuses able to work at the frontier should be tangible. Perhaps diminishing returns and declining per capita contribution to big ideas won't matter; they will become irrelevant, swamped in the aggregate.[60]

Sure, we can't keep adding researchers and creatives ad infinitum; but on the plus side, we can add a good deal more in the coming decades. There are currently around 8 million scientists in the world. By 2071 we'd need 64 million scientists to regain twentieth-century growth rates.[61] But that's well under 1 per cent of the global projected population by that point – it certainly doesn't feel unrealistic given this picture of scaling up. Indeed, it's hardly fanciful to wonder if there might not be many more. And if they can produce tools which can short-circuit the need for exponentially growing research inputs, then we are on the right path out of the problems explored in Part I.

Individuals are also better prepared. Most have better living conditions than ever before – food shortages and shocks like wars create cognitive and social impairments, for example, so the more they are removed, the more contributions are possible.[62] People are also better educated. Economists argue that in the long term, boosting human capital is the single best way of boosting innovation.[63] In 1800, 88 per cent of the global population was illiterate; by 1950 the figure had crept down to 65 per cent.[64] But since the 1980s illiteracy has plunged below 15 per cent even as the global population piled on 3 billion; the great majority now have high-school as well as primary education. We've already seen that girls in particular have benefited, and that migration flows put brains where they can have the most impact. Average years of schooling around the world increased from just 2 in 1950 to 7.2 in 2010, in better managed and resourced schools.[65]

The global university sector has flourished outside the old Western brand names, and is proving vital as the crucial incubator of new scientific, cultural and technological forms. Going into the twentieth century, universities were rare. In 1900 China, then as now the world's most populous nation, had only four universities; India, the second most populous, had five. In sub-Saharan Africa there were four and in the Ottoman Empire just one, the University of Istanbul.[66] There are now at least 17,000 universities worldwide with over 153 million students, although some estimates go as high as 207 million.[67] And universities and knowledge stores have also opened up: online resources mean anyone can access this knowledge. Among the billions until recently denied real education will be many millions of talented artists, scientists, entrepreneurs – millions, billions of flashing human imaginations emerging, fuelled and ready, into a connected world for the first time.

The pace and scale of catch-up to the frontier explodes whatever capacity we had before. And on pre-Covid trends, income could grow more than sixteen-fold over the twenty-first century, according to the OECD: an average global income of $175,000 per head in 2010s money.[68] That in turn can deliver a lot more technology, more art, more time for research and thought. Although more recent growth has slowed, the long-term performance is still astonishing. Consider the graph in Figure 2, describing two thousand years of economic history:

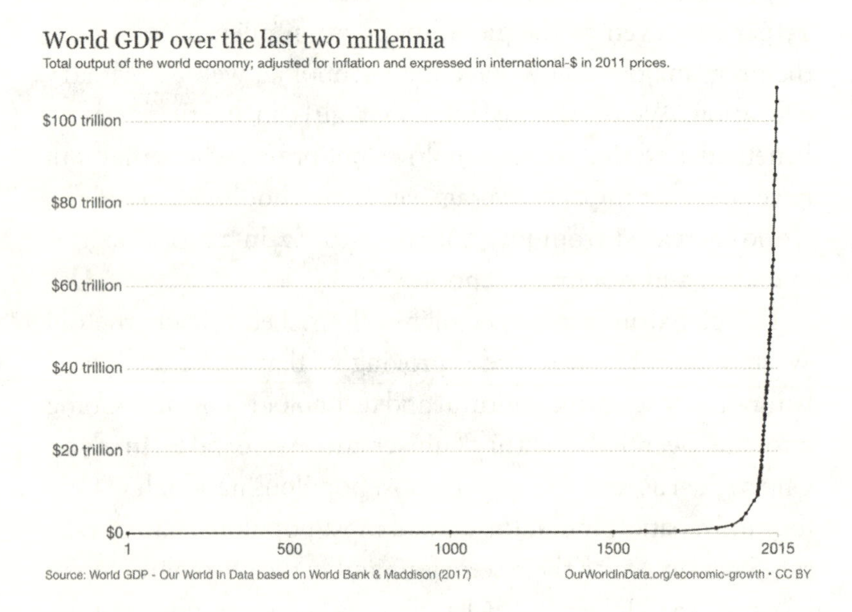

Figure 2: World GDP over the last two millennia[69]

It makes clear that slowdown is relative. The sheer scale of growth in recent years surely suggests that we have not exhausted its global impact but must still be only at the beginning of the journey. Just as over the *longue durée* the technological revolution is unfolding, so recent slowdowns in growth need to be judged against this wider, more long-term picture. Despite the slowdown, it will be surprising indeed if this picture does not mean *something* profound for the human frontier.

It should, in theory, enable an unparalleled scale-up of R&D, for example. Consider that World Bank data shows the countries with the most per capita researchers are places like Sweden, Israel and South Korea, with a little over 7000 researchers per million. Or that the current top ten spenders on R&D are the US, China, Japan, Germany, South Korea, France, the UK, Russia, Taiwan and Italy,[70] while other significant spenders include Israel, the Nordic countries, Canada, Switzerland and Singapore. Research is still concentrated in a few rich and developed nations.

The potential for catching up is extraordinary. India has only 200 researchers per million and Nigeria just 30 per million. Imagine a world where those numbers approach the global leaders. However much R&D has risen, it's heading far higher, with a very different regional balance. Judged by purchasing power parity, China is already the world's biggest economy. By 2050 India will be the second largest, Indonesia the fourth after the US.[71] Brazil, Mexico and Turkey will also grow strongly. It is not fanciful to suppose that the R&D spending league table will adjust accordingly and the totals will climb; $2.3 trillion will feel small. Consider also that, despite its highly respected Institutes of Technology, India doesn't (yet) have a single university in the global top 250. Again, what will it look like when India has dozens? Evidence shows expanded concentrations of people, connections and resource towards frontier activity have, in the past, initiated increasing returns across scientific and technical domains.[72] If this holds in future and at the macro-scale, and delivers more than just incrementalism, the Great Stagnation will be seen as a hiatus.

Look at something like Silicon Valley and it's possible to see how, initially, government investment, and latterly venture capital investment, were critical to its success: both are often weak in developing countries. In places like India or Africa the available VC is smaller, the networks much smaller, and for that reason many possible avenues are assuredly missed. Again the signs and scope for speedy catch-up are exciting. Until as late as 2015, US venture capital was more than the rest of the world put together. That is no longer true, even though it continues to grow fast; the rest of the world is growing faster.[73]

One way of thinking about the whole picture is like this: in the market for ideas we have just had the largest supply-side boost in history: the injection of more minds, resources and (as we will see) possible institutions into the mix.[74] At the same time comes a huge demand-side stimulus: hungry economies looking for new business, products, forms of entertainment, scientific and technical capacity: nations jockeying in competition and in search of ideas.

We shouldn't equate scaling up and catching up with a crude economic reductionism. Having wealth doesn't automatically translate to a better quality of thought; geo-historical progress alone is not a sufficient condition for wisdom or creativity, as thinkers as diverse as the Buddha and Ovid can testify. Yet it would be equally futile to deny the real impact of material conditions on the ability to produce ideas. Removing barriers, from hunger to the lack of basic equipment, and proactively building institutions like BGI or cultures of experimentation, makes a difference.

The Great Convergence also directly confronts the problems of the stagnant society. While no one could accuse the Indian or Chinese governments of being unbureaucratic, a lack of economic dynamism and entrepreneurship appears a phenomenon of developed economies. Although waves of populism and hype apply everywhere, as we saw with He Jiankui, different regulatory regimes may permit different rates of technological development – a source of concern, especially in the West, but an unignorable feature of this new world nonetheless.

There is a further potentially transformative aspect to the Great Convergence: a flowering of cognitive diversity. Diversity of thinking is often at the centre of innovative ideas.[75] Recall that new ideas take the form of cross-pollinations, recombinations at the crossroads of different disciplines, worldviews and objects, 'ideas having sex' in the words of Matt Ridley.[76] In a room full of scientists a novelist's perspective can be valuable, just as a scientist can give a new viewpoint in a room full of lawyers – and vice versa.

Within organisations and nations alike, thanks to the convergence, there is more likely to be a cognitive diversity of backgrounds, peoples and styles than at any previous point in history (even if there are still far too many monocultures). Spaces from labs to boardrooms were male-dominated for centuries. But studies of groups suggest that they operate more effectively when high proportions of women are present. More women working at the frontier should help deliver a burst of thinking.[77] Work on scientific discoveries indicates that the labs which integrate the most diverse scientific backgrounds and link to the widest sets of correspondence are more likely to give rise to major breakthroughs.[78] Different regulatory regimes will lead to divergent outcomes. Changing access to ideas and knowledge and equipment also creates a qualitative shift.

It's intriguing to consider how the Great Convergence might impact cognitive diversity on the largest scales. A welter of social psychology research points to important cultural differences in cognitive patterns.[79] East Asians and Westerners deploy different modes of thought. No one is saying those differences are insuperable; just that different cultures and languages lead, as you would expect, to different ways of thinking, dating back to the marked differences epitomised by, say, Aristotle and Confucius.

Westerners broadly focus on objects and categorisations, looking at single agents and formal rules of logic. Asians are more attentive to context, contradiction and webs of interrelation, thinking characteristic of traditions that look for harmony and towards the collective. The West tends towards individualism. Western children learn more nouns – objects – whereas Eastern children learn more verbs. This leads to a differing sense of causal relationships; the West has broadly followed a clearly identifiable atomistic single cause model, while Asians look to complex, context-dependent webs of causality.

When looking at an image of a fish tank, Westerners focus on the individual fish, Asians look at the whole context. Parents play with their children in different ways – Western parents focus on objects, toys, whereas Japanese parents focus on the relationships between themselves and the toys. Farmers growing rice, with the complex irrigation and terracing involved, need a very different mindset from wheat farmers; the former need to be more 'collectivist', against the more individualist wheat farmers.[80]

The meeting of thought processes is already creating transformations – remixings of spiritual or medical traditions. Asians joining US-style evangelical mega churches; Americans practising tai chi and acupuncture. It implies that different styles of argument and debate, different business practices, legal processes, medical considerations, modes of enquiry, political imperatives and contractual relations will all have their place, and hints that the spread of ideas will not simply result in pale simulacra. Diversity, cognitive or otherwise, is now an inescapable feature of the world, experienced in our offices, studios and labs, at home and in digital media as well as on the scale of continents. This is superfast cross-pollination on a civilisational level.

History also suggests that bursts of creativity or original technology tend to be highly concentrated in place and time. This means a diverse scale-up is vital to avoid an entrenched and lasting slowdown in our big ideas. Cardwell's Law suggests that any given society, left to its own devices, will tend towards stagnation.[81] However, even as fecund moments burn out in one country, they can be relit elsewhere. From the late sixteenth century to the late nineteenth the frontier passed from Italy to the Netherlands, to the British Isles, to Germany and then to the US. In each the embers of innovation first burned bright before dimming, but there was always somewhere else to carry the torch. Overcoming Cardwell's Law requires options, meaningful frontier competition. In early modern and modern Europe the fragmented political, religious and intellectual environment offered that. Unlike in say homogenous imperial China, heterodox thought could flourish in the margins – there was likely succour for a rebel idea, somewhere for the torch to go.

Back in the twenty-first century, scaling up suggests that the torch can still be passed. Competition is alive and well; the complacent, unipolar post-Cold War order is fractured. Take away the competitive pressure, the diversity at the frontier, and we'd face the fate common to sprawling empires: stagnation, decline, diminishing returns. The uncomfortable truth is that, from the perspective of ideas, scaling up was probably the only option to keep us moving forward.

There is no automatic mechanism here, no necessary overcoming of the stagnant society's idea paradox, just greater probabilities that new ideas will arise. Negative shocks, like the coronavirus or worse, could make immense dents in society. Indeed, an event similar to Covid-19 might not only halt the Great Convergence but lead to a wholesale retreat. Moreover, much of the world that should contribute in the twenty-first century is acutely vulnerable to another mega-trend: climate change. Sub-Saharan Africa, the Middle East and India could be subject to devastating heatwaves and droughts. Global metropolises from London to Jakarta could be swamped by rising seas – so much for those crucibles of cognitive diversity and invention.

China's embrace of redoubled authoritarianism, its blanket surveillance and the ideological purity of Xi's Communist Party does not bode well for the kind of risky experimentation and free communication associated with new ideas. Indeed, they could suffocate work at the frontier in the name of stability and orthodoxy, even as opportunities for economic growth slow, potential allies and collaborations are spurned and its population rapidly ages. There are few modern precedents for non-open societies really delivering at the frontier. Xi's gamble is that China can achieve a new generation of ideas without the accoutrements of liberal democracy or allowing those ideas to develop in unpredictable ways. But this could backfire; after all, Ming Dynasty China closed itself and fell back from the cutting edge. My hunch is that China's model will prove effective at tackling well defined problems, giving purchase to already conceived and partially executed ideas, exactly that portion stuttering in the West. But over the long term it could struggle to conceive and execute the radically new.

Suffice to say, it is not settled whether a society that reins in its entrepreneurs and silences critique can reliably disrupt itself in the sense used here. Moreover a war, or even a simmering conflict, between China and the West, which seems more probable all the time, might stir some big ideas but could be catastrophic beyond imagining.

Nor may China present such an optimistic economic model. While it has exemplified catch-up growth, other countries may not have the scale or the opportunity to simply copy it; as robots and sticky supply chains leave production where it is, growth for those countries may follow a more winding path.[82] Ideological and bureaucratic pressures, stifling autocracy and corrosive kleptocracy are hardly absent in emerging economies elsewhere, for that matter. Should any of those difficulties increase, the freedom for new thinking could be narrowed sufficiently to outweigh the positives. And behind it all, He Jiankui is hardly a comforting or hopeful example for where the future might lead.

However, even despite this, thanks to the increase in global connectivity and the gains that have accrued, over the next years the break since the 1970s and 80s will bed in. Looking at the past two decades we find only the shoots; look to the coming decades and beyond, we will find the full fruit.

This will be the last such burst. Once we are all at the frontier, there will be no more quick wins in terms of connecting or adding or educating people. As we saw in Chapter 5, eventually the cross-pollination of cultural interaction will exhaust itself. By the century's end, population will decline – perhaps precipitously. China alone could lose up to 600 million people.[83] The Great Convergence is, in this sense, a last hurrah, the final crescendo of a process that has been gathering pace for hundreds of years. If it doesn't bear fruit, then one question will be answered: diminishing returns and the breakthrough problem are here to stay.

But we can't say that yet: the Great Convergence is long and deep, and is only just making itself felt. We have reasons for optimism.

The Institution Revolution

There can be few people more emblematic of heroic individual invention than the two who brought electricity to the masses: Thomas Edison and Nikola Tesla. Today each has a substantial cult of personality: creators of the modern world, synonyms for big ideas. Their personal interactions and rivalry in the 'current war', where Edison's direct current slugged it out against Tesla's alternating current, only add spice to the drama.

But of course this is a myth. In fact both of them show how much, by this point in history, invention was an institutional affair.

Edison cultivated the impression of a lone genius, growing up poor in rural Ohio, constantly experimenting and tinkering and selling newspapers on the Grand Trunk Road to Detroit. In his early days, working on telegraphic innovations in Boston, he was pretty much a one-man band. But his later success depended on producing one of the world's first and most famous industrial R&D labs. When he was born in 1847 it was still normal for thinkers and innovators to work alone; by the time he died in 1931 this new form dominated invention.

In his Menlo Park lab, Edison worked closely with at first around fifteen others, and eventually up to two hundred, in what he called 'the invention business'. These engineers or 'muckers' worked in a single large room strewn with the detritus of technological innovation. Hours were hard and long. The atmosphere was part lab, part machine shop, part frat house. The muckers were instrumental: one, Charles Batchelor, was particularly important, a skilled draughtsman and mechanic whose hands-on approach complemented Edison's so much he was even awarded 50 per cent on their co-inventions.

Work was divided between research for clients in fields like telegraphy, electricity or the railways, and proprietary R&D. But the lines were never clear. Edison liked to claim a freewheeling approach, but Menlo Park was meticulous and methodical. In finding the right plant material for the first carbon filaments, the team tried no fewer than 6000 types. When developing the nickel-iron battery, they performed 50,000 experiments.

Between 1876 and 1881 alone, on a workbench that ran the length of the New Jersey lab, there came a series of astonishing inventions and innovations, from vacuum pumps to filaments and lightbulbs, generators, mimeographs, voltmeters, the phonograph, a whole array of improved telegraphs and telephones. In six years the factory landed over 400 patents. It became a byword for invention. All this cost money: Edison was bankrolled by the preeminent financier of the era, J.P. Morgan, without whom he would have struggled to survive.

The Serbian-born inventor and sometime collaborator and rival of Edison, Nikola Tesla, has established a reputation as a lone genius, a singular savant big-idea generator almost without peer: the man behind the AC motor and the building blocks of radio, but also everything from X-rays and remote control to robotics, microwaves and TV. Thanks to the film *The Prestige*, the popular image of Tesla is of a man huddled away in the Colorado wilderness single-handedly revolutionising technology and science. Again, he cultivated this image, writing: 'The mind is sharper and keener in seclusion and uninterrupted solitude. No big laboratory is needed in which to think. Originality thrives in seclusion free of outside influences beating upon us to cripple the creative mind. Be alone, that is the secret of invention; be alone, that is when ideas are born.'

No doubt Tesla's methods often seem to approximate those

of the solo inventor: performing feats of visionary imagination and calculation in his mind – even as a boy, after seeing a picture of the Niagara Falls, he envisaged a great wheel generating electricity from its flow. But even Tesla required assistants, labs and tools. Even Tesla spent time chasing financiers, Morgan among them, but also others like John Jacob Astor. And in the current war with Edison, it was the entrepreneurial vim of George Westinghouse that pushed out the network, who bought in Siemens generators, who went toe to toe with Edison. By the time Tesla got to work, Westinghouse was already in hot pursuit, risking everything on his AC gamble.

Eventually Westinghouse's gamble paid off. Having endured a bruising PR battle and almost gone out of business in the early 1890s, Westinghouse and Tesla won the contract to light the Columbian Exposition in Chicago in 1893; AC, with its ability to transmit over long distances, won the war. At the same time, Morgan began turning Edison Electric into General Electric. There is, though, another twist. Tesla may have helped win the current war, but he died an overlooked and eccentric old crank. Edison died exemplifying invention, having built one of the world's most dynamic labs and largest companies.

He'd lost the current war but, despite himself, won the battle of how invention and innovation would be pursued in the twentieth century. Large, well-organised laboratories had become central to the process of invention, innovation and ideation. Edison's way of inventing – his real way of inventing – became the norm. Behind his lightbulb, generators, phonograph and Kinetoscope was an institutional method of invention. What he'd built was a miniature factory for ideas: in Edison's words it was for 'rapid and cheap development of an invention', nothing less than 'a minor invention every ten days and a big thing every six months or so'. It was precisely the system, the network, the organisation, that had become significant.

In the early nineteenth century a painter like Samuel Morse could still invent the telegraph. Lone scribblers and tinkerers were common – English reverends like Thomas Bayes, gentleman thinkers like Darwin, renegades and loners from Nietzsche to Marconi. But while inventions like the telephone, radio and the motor car were still associated with individuals (even if in reality many individuals were about to realise them), each had quickly led to the growth of behemoths like Bell Corporation, RCA or General Motors. Reaching purchase required more resources than ever. The 1IR had created the factory; the 2IR institutionalised every aspect of ideation.

Prior to the Manhattan Project, the world's biggest research project had been the German chemicals conglomerate BASF's effort to produce the Haber-Bosch process that fixed nitrogen from the air as a fertiliser: a process without which the modern world would be unthinkable.[84] As fields developed and complexity increased, it became rare for scientists to operate outside a university or corporate research laboratory, just as the focus for entrepreneurs became institution-building itself. Electricity was a particular catalyst of organisation density: the number of engineers in America grew from 45,000 in 1900 to 230,000 in 1930, over 90 per cent of them working for industry.[85] To invent and then scale up required time, money and people that individuals couldn't match.

Thomas Edison here becomes a historic pivot: a brilliant one-man powerhouse of invention who also, with his corporate lab, established the mould for a more institutionalised future. Earlier in the book we saw the rise of big science, which parallels this story. In the early twentieth century, Marie Curie could conduct her experiments almost single-handed in a small

Latin Quarter lab; just decades later, nuclear physics would require some of the largest, best funded and most complex scientific organisations in history. As late as the 1950s individual authors or inventors were still capable of writing that world-changing paper or making a brilliant invention; but, as we saw in Chapter 5, in the twenty-first century these are the province of ever-expanding teams.[86] If ideas have always rested on networks and institutions, over time that process has been formalised and expanded. Edison and Tesla may indeed be heroic individuals, but they are also exemplary of how big ideas rest on an institutional foundation.

Cognition is institutional. In the words of the anthropologist Mary Douglas, 'thinking depends on institutions'.[87] No surprise then that the history of ideas is also the history of institutional growth.[88] Libraries to store books and knowledge; laboratories to conduct experiments; hospitals to cure people; parliaments and think tanks to form policy; businesses to create products and services; studios to channel and cohere creativity. Institutions provide resources, norms and platforms for work; they create areas of common knowledge, practice and understanding; they own and build the necessary tools and materials. Now economists and organisation theorists are starting to quantify how institutions amplify and aid discovery, especially in the context of great accumulations of knowledge and complexity.[89]

Institutions offer mental frameworks as well as resources and colleagues. They tell us what matters, what we should do and why, and then provide the ability to act on that. Institutions store, transform and distribute knowledge. They learn with experience. Indeed, organisations exhibit qualities of intelligence: they monitor what happens around them and react, have goals, self-awareness and the ability to integrate these functions in complex parallel processes.[90] All of which means the design, structure, endowment and number of different institutions and organisations powerfully influence how ideas are produced.

Even going back to the ancient world, institutions were at the heart of the search for ideas. Think of how the more open, deliberative polity and metaphysics of the ancient Greeks led to a litany of big thinking, from Thales and Democritus to Plato and Hippocrates. It also shaped specific organisations: while Socrates contented himself with the agora, Plato founded the Academy and Aristotle the Lyceum. Archimedes was conversant with the great Library of Alexandria, the beating heart of scholarship in the classical world. Even before that, at the earliest dawn of civilisation in Mesopotamia the needs of bureaucratic and financial institutions drove the development of writing and maths.

Individual thinkers have always been tempered by great institutions like religions, and within that specific organisational forms like monasteries. But for much of history such institutions stifled big ideas outside a narrow realm. Nonetheless, big ideas need social norms and structures that support their creation and dissemination, and alongside the Scientific Revolution's new toolkit came structures promoting the exploration and dissemination of original thinking. There was, as we have seen, a Republic of Letters; Europe's thinkers and writers in a constant dialogue, a giant and dispersed peer review mechanism, communications platform and mentorship scheme. This amorphous institution was underpinned by the bracing new values we saw spark an ideas revolution in Chapter 1: delight in novelty, scepticism towards ancient knowledge and a focus on 'usefulness' and 'improvement'.

Slowly but surely came the growth of more specific institutions devoted to this revolution. Some of the old scholastic universities started to adapt: places like the University of Padua hosted Galileo and Vesalius and had William Harvey and Copernicus as graduates. On the Danish island of Hven, the astronomer Tycho Brahe built the Uraniborg, something like the world's first modern research and experimental institute. By the latter half of the seventeenth century, organisations like these are increasingly important supporting structures for ideation; the founding of the Royal Society in 1660 is a key moment, just as mercantile ventures started to form joint stock corporations, artists collected themselves into schools and key forms of communication like the newspaper, the scholarly journal and the novel took shape.

In Britain the eighteenth and nineteenth centuries saw the growth of an 'associational society' around knowledge, extending from the Royal Society to other formal (and often regal) institutions dedicated to knowledge and ideas: the Society of Arts, Manufactures and Commerce, the Royal Academy of Arts, the Royal Institution, and a wave of universities, informal debating clubs, public libraries, mechanics' institutes, professional societies and more specialised learned bodies (from the Archaeological Association, formed in 1843, to the Zoological Society of London, established a few years before in 1826). These were mirrored around Europe, in for example Frederick the Great's reformed Prussian Academy of Science. In the seventeenth century learned societies were in their infancy; by 1880 London had 118.[91] Universities continued their march: the Scottish universities played an integral part in the Scottish Enlightenment, while later German reforms saw their universities change into research powerhouses, followed by those of the United States. This institutional shift had a concrete impact on the delivery of ideas: the number of books published rose steeply, for example, and a survey of British patents granted between 1752 and 1852 shows that the more 'knowledge access institutions' like the above there were in a given area, the more patents it generated over the next decade, the higher the impact of those patents and the more exhibitors and prizewinners it had at the 1851 Great Exhibition.[92]

Even in the post-Romantic world of visual art, where individual creative acts held primacy, turning points like the *Salons des Refusés*, instrumental in creating a movement like Impressionism and a rebellious mode for art, were collective efforts. The isolated, paint-spattered garret tells only half the story. Artists may work alone but they do not create alone. Around them grew an institutional architecture of colleagues and fellow travellers, materials suppliers, dealers and gallerists, training schools and critical outlets, and behind it a market for art, an economy of prestige and distinction, and of ideologies of novelty and revolution.

Over centuries, a virtuous circle set in. Cultures and values began promoting cutting-edge work, freedom of enquiry, peer review, attitudes of experiment or improvement, and so on. These values enabled the creation of formal institutions.[93] Their success in turn buttressed that culture. Over time an effective institutional and cultural matrix dedicated to discovery and creation evolved. By the twentieth century it was rare indeed for big ideas to come without any organisational context – even novelists and musicians have publishers, record labels, informal networks and structures of critical appraisal. At this point we reach Edison, Menlo Park and all the rest of it: ideas generated, as a rule, in and through institutional contexts.

A society's ability to generate significant ideas hence rests on its institutional base; and as ideas require more people and resources, so they rest more heavily on that base. The context of scaling up gives reasons for optimism here. Many institutional basics are recognised worldwide: quality control, a premium on new knowledge, an understanding that deference to old authority is inhibitory, the need for basic resources. We're beginning to grapple with the nature of collective intelligence as the primary means of large-scale action in a complex world. Above all, a scaled-up world proffers more and better equipped organisations, another supply-side boost in the quest for ideas.

Big ideas are low on the agenda of most organisations. Many, perhaps most, institutions narrow thinking; even when they claim the opposite they embrace the pressures of stagnation. But there is a species devoted to their production. Thanks to the Great Convergence, they are spreading, seen as essential for development, competitive advantages for nations, cities and individuals. I call them breakthrough organisations: institutional forms specifically directed towards big ideas. They are rare and fragile, but the best examples, like Google X or the Cavendish Laboratory, are recognised and can expand quickly. Full of high-calibre people, the space and freedom they afford gives them nonlinear impact: inventing a new genre, uncovering the structure of the atom, building the first computers. Much of the future of ideas will hinge on the twenty-first century's ability to nurture and create these organisations in sufficient health and quantity.

The scale of breakthrough organisations like Alphabet or the University of Cambridge or Disney or the Indian Ministry of Science and Technology, would, in their domains, be equivalent to that of the entire world not too long ago. They experience the pressures of the stagnant society but they are also the central mechanism by which we create space insulated from its deleterious impact. The pockets I mentioned in Chapter 6 are in large part organisations like this that create the necessary structures of incentives, regulation, intellectual space and material resource. As the inhibiting trends we saw in previous chapters intensify, they are balanced by the growth of places like the Beijing Genomics Institute or the Indian

Space Research Organisation. Global scale-up implies that, opposite Alphabet and Cambridge, Tencent and Tsinghua will create their own substantial pockets. Weighting the global balance sheet with more and more capable organisations means carving out more pockets for ideas. There is no reason elite Latin American or African breakthrough organisations will not vie at the frontier with American universities or Chinese tech companies. Innovation, invention and ideas have been concentrated in just a handful of cities and lauded institutions. Even a relatively narrow opening – Shenzhen *and* Silicon Valley, Mumbai *and* New York – represents a revolution.

I think about three varieties of these pocket-creating organisations. First are what I call imaginative networks: loose and creative assemblages collectively imagining the new. They are perhaps the oldest and most natural forms of breakthrough organisation, often small and tight knit: say, the two Wright brothers and their interlocutors. They may be more clubbish, like the Lunar Society of Birmingham, a gathering instrumental in the early Industrial Revolution, and range from the Founding Fathers to the Homebrew Computer Club. Brian Eno calls it 'scenius . . . the communal form of the concept of the genius': localised efflorescences in places like Athens, Florence, Edinburgh, Vienna or the Bay Area.[94] Such networks are central to the creative process; from the Impressionists to the Inklings, punks to Bauhaus, creative bursts happen within groups where ideas are bounced around.

Sometimes these networks are grounded in a physical location – at the Library of Alexandria or Studio 54 and Warhol's Factory. Today they are recreated within some companies – think of Pixar's Braintrust or the writers' rooms on the best TV shows. Big ideas emerge from these networks so often that a history of ideas, from the Vienna Circle to the Frankfurt School, overlaps with astonishing precision.

Second come experimental spaces, more formal institutions specifically designed for high-risk trial and error. The industrial labs we saw earlier are good examples: Bell Labs or Xerox PARC or Lockheed's Skunkworks, Apple's Design Team or Nike's Innovation Kitchen. The world's great laboratories also count: the MRC Laboratory of Molecular Biology, with its twelve Nobel Prizes, CERN or Lawrence Livermore National Laboratory. So too would a range of institutes supporting diversity of innovative thinking – anything from Princeton's Institute of Advanced Study to the MIT Media Lab, the Santa Fe Institute and Japan's RIKEN and the eponymous institutes of Max Planck, Louis Pasteur and Francis Crick. But the diversity is greater still – you could argue that the structure of Oxbridge colleges enables experimental spaces, just as organisations like Y Combinator do for startup ideas, or ARPA does for technology in general.

Lastly come even purer breakthrough organisations – those with a specific mission geared around executing a particular idea. Places like Bletchley Park or the Manhattan Project, which grew out of the Second World War, are emblematic here – faced with a monumental problem or challenge, they assemble a taskforce to achieve it. This form may be the only way of taking on many of our biggest questions. DeepMind, the Human Genome Project and IBM's quantum computing centre resemble this kind of organisation, with a clear goal and the means to achieve it. There are perhaps fewer breakthrough organisations than one might expect for a society that has reached this point, but they still exist: SpaceX focused on getting to Mars, SETI finding aliens, ITER chasing fusion.

As time goes on, thanks to advances in management and organisational theory and the understanding of innovation, we should – at least on paper – be better able to design, build and maintain breakthrough organisations, including experimental new forms of organisation.[95] Many social pressures, skewed metrics and misaligned incentives can then be ignored: researchers in a wealthy tech company worry less about citations and their next grant, fellows at a well-funded experimental art college worry less about champagne-swilling art collectors, and so on. There could also be a silver lining to the existence of larger teams. Patent records show that knowledge diversity, present in the best teams, is helpful to breakthrough invention.[96] One study suggests that despite the downsides we have seen, bigger teams, and especially big and diverse teams, can better weed out bad ideas and wrong avenues to focus on what will really matter.[97] Like larger civilisations, they throw up more interesting and novel combinations and outlier ideas. Organisations that support larger teams will have more costs and pressures, but will also see more benefits.[98]

Tools like the Internet should make imaginative networks more powerful and easier to form. Even if we have grounds to be disappointed in this regard, tools from Kaggle to Wikipedia suggest that it can create new forms of breakthrough organisation. Resources should be available, amassing behind the missions of breakthrough organisations. Experimental spaces proliferate around the world, sucking in talent.

Despite the pressures of stagnation, signs of institutional resilience and growth are apparent: the British government has belatedly recognised once again the importance of blue skies research and established a kind of UK-focused ARPA (called ARIA); venture capital has slowly turned back to so-called 'deep tech'; funding bodies like the Wellcome Trust are ring-fencing money for the riskiest avenues of research. The productivity slowdown isn't uniform: at the top of the distribution a small batch of super-high productivity companies keep innovating – and growing. Covid-19 showed, if it were needed, the value of strong and free research spaces and organisations, places like the Jenner Institute or startups like BioNTech. Even if the overall impact of the pandemic veers towards the conservative and oligarchic, there should be lingering leeway given to places that could make a critical difference in future.

Yet, as time goes on, breakthrough organisations can become more subject to the many pressures of stagnation: the whims of shareholders, the demands of conflicting bureaucracies, political pressures both internal and external, the fragmentation, distraction and polarisation of social media. Even some of the most optimistic observers about scientific and technical progress cite risks to the institutional mix as the most 'serious concern' for its continued success.[99] Just as there is a tension between the difficulty of ideas and the tools, techniques and means to engage with them, so there is a tension between institutions and the pressures placed upon them.

The most pressing question for the future of ideas is whether the scaling-up effect and its great accumulations and agglomerations of knowledge, people and capital can produce sufficient pockets of innovation for the next generation of big ideas. That future is caught between the capacity of this scaled-up world to create breakthrough organisations at an expanding pace and our relentless capacity to stifle innovation in the face of complex and challenging ideas. We'll return to this tension in the next chapter.

My answer is a tentative yes: we can create those pockets. This civilisational shift does indeed mark a break. But only if we consciously push it. The Great Convergence means that building those organisations is absolutely possible; the real question is one of will.

*

Institutions and organisations are 'collective brains' or 'superminds'.[100] One mind alone could not have invented Palaeolithic tools or food preparation methods, let alone run a modern country, fabricate millions of iPhones or invent CRISPR.[101] Only orchestrated collections of minds could do so. Almost all human activity is underwritten by an intricate institutional mesh that enables this. The physical weight of everything humans have developed – buildings, roads, cars, etc. – is an estimated 30 trillion tons.[102] Those 30 trillion tons rest on an ephemeral institutional foundation, the products of superminds.

This mesh is at a turning point. We've already seen that size and interconnection are important for the generation of ideas. On both counts, our societal operating system runs at record levels. That maturity and size could alone make a qualitative difference. Think of the human brain: individually its 86 billion neurons are not capable of all that much, but collectively they produce consciousness and intelligence. The number of neurons within a brain is a good species-level predictor of intelligence, just as the more transistors there are on a chip, the more tasks it can accomplish. Scale matters, but so does organisation: the power of the human brain comes from its dense web of neuronal connections. Adam Smith's pin factory is so important because such changes in organisation lead to rapid improvements in productivity: one person alone can produce a few pins, but more people, organised in the right way, can produce exponentially more.

Scale and co-ordination produce a new kind of intelligence and a different quality of thought. If that is also true for human organisations, the present moment is especially significant.

There are currently 8 billion humans on the planet, with more opportunity than ever. The tools stitching them together are vast: at least 15 billion machines, soon to be many more;

trillions upon trillions of transistors networked together and through all those brains and neurons.[103] Human and machine now live in complex symbiosis, interwoven into patterns of machine–human collaboration. Consider that even something as basic as going to dinner with friends is now a complex dance of technology and biology: text messages, synced calendars, food ordered via the Internet but grown and assembled from around the world, co-ordinated to one kitchen through an array of markets and technological interfaces.

Just as the number of and density of connections in the human brain make it so potent compared to that of an earthworm, this picture of both scale and connectivity should, I think, lead one to wonder what it means for humanity.

We could already be at or on the cusp of a higher level of collective intelligence, a new global mind, an interconnected collective brain pulsating through our myriad interactions, stores of knowledge, flashing neurons and transistors, linked through the dense and powerful institutional and fibre-optic network of our age. Gaia Vince believes that interplays and levels of culture, biology and altered environment render humanity into a new kind of superorganism.[104] She calls this *Homo omnis* or Homni – a shared resource stitching us into one closely interrelated world network, a mash-up of gene and machine, human minds and outsourced technological and institutional systems. Homni is the emergent property of all this activity, working like a scaled-up brain. A century ago H.G. Wells called it the world brain; Pierre Teilhard de Chardin the noosphere. Kevin Kelly calls it the holos, a new planetary layer composed of billions of human minds, chips and artificial intelligences. We are linked into a single, singular entity and should expect further phase transitions to result, transitions that could unlock, like the most elaborate tools, a set of big ideas that changes everything:

> This convergence will be recognized as the largest, most complex, and most surprising event on the planet up until this time. Braiding nerves out of glass, copper, and airy radio waves, our species will began wiring up all regions, all processes, all people, all artifacts, all sensors, all facts and notions into a grand network of hitherto unimagined complexity. From the embryonic net was born a collaborative interface for our civilization, a sensing, cognitive apparatus with power that exceeded any previous invention.[105]

This, the holos, is the breakthrough organisation to end them all.

The Great Convergence is, then, a bigger process than just economic catch-up or the proliferation of smartphones. It encompasses all of that and implies something more: a phase transition in the human mind itself. Perhaps we don't need to posit an artificial superintelligence or augmented minds because something like them is already here. The Great Convergence is beyond any country or organisation; it is truly a planetary moment, an interweaving of people and tools on a massive scale and in an immediate timeframe.

Whether or not you believe it forms a global mind, the plain fact of this scaling up and what it could mean for ideas is extraordinary. Previous step changes in the collective brain led, among other things, to democratic governance, modern science and the Industrial Revolution.[106] This one is still very much in process. Facing off against the challenge of the Idea Paradox is a startling society-level emergent entity. If anything can ascend the ladder, overcome the breakthrough problem, swamp or reverse diminishing returns, it's this.

9

The Uncertain Horizon

Three Trajectories

The future of big ideas hangs in the balance. On the one hand, the inexorable ratcheting up of difficulty, complexity and paths already taken and the airless purgatory of stagnation; on the other, the promise of a new toolkit, and the crescendo of a scaled-up, joined-up world fizzing at the frontier. Their tectonic interplay is radically uncertain, but the future of human thought rests between these forces.

Our best bet for understanding that future lies in finding the most plausible stories for what happens next, in thought experiments as much as forecasts. Predictions are usually wrong, sometimes very wrong indeed; open-ended terrain like this is better served by thinking alert to the various uncertainties in play. The sketching of scenarios and trajectories deepens our understanding, and offers a mental tool for thinking across a contradictory and protean landscape.[1]

The future of big ideas is also still up for grabs. This chapter addresses the question of what practical steps, policies, attitudes or projects might help. But first it looks at three possible trajectories in more detail.

The Long Twilight

Taking progress for granted is a mug's game. Historical precedents abound of civilisations which, having reached the forefront of knowledge, then fall apart. The story of the Roman Empire is instructive in this regard, but not for the reasons you might think. According to the classic account of the fall of Rome, the Empire became increasingly decadent and dysfunctional, making it easier prey for Northern barbarians who sacked the place; thus ending the classical era and ushering in what was once known as the Dark Ages (and now more properly as Late Antiquity). However revisionist you like your history, there was an obvious rupture. What this account misses is how, in intellectual terms, the 'fall' came earlier, and for very different reasons. Blame Jesus as much as feckless emperors.

Before the fall came an epic 'closing of the Western mind' that reverberated down the centuries.[2] Until the fourth century CE, Rome had (just about) kept the Hellenistic torch burning, a culture of enquiry clinging on to intellectual openness: the astronomy of Ptolemy, the medical insights of Galen, the philosophy of Plotinus. A society where reason and questioning had its place. Then came the Emperor Constantine's conversion to Christianity in 368 CE.

A Christian tradition had long before sprung up in opposition to the Greeks; St Paul in particular had attacked their philosophers. They were, he thought, doomed, the whole rationalist project heretical. After Constantine, his new faith-based ideology was in charge, and it would not tolerate a reversal. Books by newly proclaimed heretics were banned, punishable by death. A Neoplatonist mathematician and philosopher like Hypatia of Alexandria was dragged through the streets, torn to pieces and burnt by a Christian mob. Her death represents, in the words of one historian, 'the end of the era of Greek mathematics'.[3] The Library of Alexandria is popularly supposed to have gone up in a great conflagration; in fact, it seems, it slowly crumbled, ignored and neglected over long years as learning was abandoned.[4] Vast numbers of classical texts were lost forever over this period, deliberately destroyed; what we have left is only a snapshot of the textual heritage of the classical era. Plato's Academy held on for nine hundred years. But faced with the absolutism of the Church under Justinian it eventually closed in 532, its last bedraggled philosophers fleeing to Persia.

The tolerance, pluralism and supple energy of the Greek legacy was crushed under a dogmatic steamroller. What had been a fringe cult grew into a populist movement commandeering the levers of power and shutting down channels for new thinking. While the early church was given patronage and tax exemptions, secular authorities tightened control via the imposition of its rigid orthodoxy.

As A.N. Whitehead put it in *Science and the Modern World*, 'In the year 1500 Europe knew less than Archimedes who died in the year 212 BC.'[5] It is extraordinary that science and philosophy had hit a peak even before Rome, when Aristarchus hit on the heliocentric model, Pythagoras, Euclid and Archimedes pushed mathematics forward. For century upon century Europe now went down a path of regression. Even in its moments of relative renaissance it saw Aristotle's static universe or a world of perfect heavenly spheres populated by angels. That it was so persistent and total over such a long time should surely give anyone trying to understand human ideas pause for thought.[6]

You might confidently predict that a fraying empire

struggling to pay its jaded and underperforming legions would buckle. It's harder to spot how internal factors undermine your own foundations when you are living through such a turning point; or when, like Tokugawa Japan, a nation turns its back on the world. There are other ways you can fall, of course – civilisations have repeatedly come to grief via external shocks like famine, disease, or invasion from the steppes or across the seas. We might be prey to any of these, even if the tools we have to combat them are so much better. For me the Roman example is so instructive because it's subtle, slow and insidious, and began with a shifting intellectual and cultural practice: the rise of one big idea that snuffed out all the others.

Who could calmly say that something similar is impossible in the twenty-first century? Recent examples of extremist one-note thinking have burnt bright but withered fast – think the Cultural Revolution, the Khmer Rouge and Islamic State, even if regimes like that in North Korea cling on. But the lesson is that such events can occur slowly, and to imagine they wouldn't happen here, that we are immune, is naive. Recent years show how vulnerable societies still are to the demagogue, the ideological kneejerk, the closure of porous spaces, the intolerance of dissent. Twenty-first century societies are not the Romans. But nor have they completely left them behind. If any of the negative forces currently evident gather pace, it could be us.

So, what might a similar turn look like today? Dystopian fiction is helpfully devoted to imagining answers. In the worst case there are plenty of outlines of a post-apocalyptic world, as seen in *Mad Max* or *The Road*. In the event that some huge shock undermines the entire infrastructure of modernity, it's safe to say the frontier will be retarded by centuries. Survival, the basic collection of resources and blood-soaked tribal politics will dominate a second dark age.

Ian Morris's analysis, as we saw in the last chapter, shows that civilisations do hit ceilings in their social development which they cannot break through.[7] This happened around 1100 BC, and then again to the Romans and Song Dynasty China. A separate survey of sixty civilisations suggests that their average duration is a little over four hundred years; the Romans are no outliers in falling apart.[8] Only with the advent of the 1IR did this pattern seem to end, but what if we hit the next ceiling in the coming decades? Coupled with the negative forces, shocks like climate change, pandemics or resource depletion look like classic ceiling events, where the nature of a civilisation's development bites back. Those worries of collapse we saw in Chapter 1 haunt society. The patterns of history seem to be breaking down, but it would be unwise to bank on that.

There is an even darker interpretation of a downward spiral: the Fermi Paradox. Lunching with colleagues at the Los Alamos National Laboratory in 1950, the great physicist Enrico Fermi was chatting about aliens, as great physicists do. 'But where are they?' he asked. Given the size of the galaxy, he reasoned, an alien society should have created a space-faring galaxy-spanning civilisation. The Drake equation suggests there could be thousands of such civilisations; in 1967 Joseph Shklovsky and Carl Sagan suggested there 'should' be around one million alien societies around the Milky Way.[9] How strange, how disturbing even, to have heard nothing. Perhaps there is some universal law that life tends to destroy itself; a 'Great Filter', some process of destruction or self-destruction, that occurs again and again between the genesis of the first multicellular organisms and the development of a pan-galactic super-civilisation. The theory maintains a haunting pull on the imagination: that our ideas will never alter our trajectory into inevitable oblivion.[10]

Say we avoid an existential threat but get knocked back, hitting a hard ceiling or falling prey to a black swan, society-shattering event. Eventually this might settle into a steady state, unlikely to be a fertile ground for ideas: in Robert Harris's novel *The Second Sleep*, a post-crash society eventually reverts to the medieval, deeply religious and conservative; still, centuries on, blaming the freewheeling intellectual, cultural and technological climate of the earlier era for the downfall and chaos that ensued. Stability takes primacy over novelty; blunt belief over truth.

Should it happen, it will (probably) take us all by surprise. Meanwhile there are more pressing dystopias. You can, for example, see different lineaments of populism run amok in two classic dystopias, *Fahrenheit 451* and *Brave New World*. In the former, books are burned by 'firemen' on the orders of an American government intent on stamping out free thought. In the latter, society is too busily intent on its own drug-induced pleasure to bother with thinking at all. Some creeping combination of the two remains plausible. AI and biotech might be exciting tools, but it's also conceivable they will bring out the pitchforks. Or maybe VR really will be better than real life, and that's that.

In a more mundane scenario, research productivity keeps falling. Difficulties indefinitely delay new tools. Big new ideas become unaffordable. Culture ossifies. Political experiment becomes too risky. Population shrinks. The global economy edges towards zero sum even as climate change hits. Perhaps in such a scenario the realm of ideas would become something akin to Herman Hesse's classic *The Glass Bead Game*: knowledge and culture reduced to an arcane, almost infinitely intricate and rarefied yet ultimately pointless game for a cloistered elite.

This is the best of the bad outcomes we can hope for: a long twilight rather than a final conflagration. All of this, though, rests on a big assumption: that the new tools and the scaled-up world are not enough to rescue us from stagnation.

The New Utopia

What if things go right? All the ingredients are in place for the most astonishing series of further transformations. There is, theoretically, no reason why we can't see lifestyle improvements equalling those of recent centuries or find answers to any remaining questions. And yet dystopian visions dominate the contemporary imagination. The tenebrescent tone of twenty-first century future-gazing has not gone unnoticed.[11] As the breakthrough engine stumbled, we stopped dreaming of those shining post-war utopias. More interesting and, in an era of mounting problems, pertinent to dwell on how things fall apart.

But as with the long twilight, there are clear precedents for steep upturns in our capacity for new thinking. How about this for a historical thought experiment: if you were an alien visiting Earth in the early sixteenth century, which country would you most expect to spark a scientific and industrial revolution?[12] You would not pick England. The obvious candidates were China or India, with their vast populations, powerful militaries, massive economies, developed cultures of administration, science, technology and learning. Ming Beijing and Ottoman Constantinople were both ten times the size of London. The Chinese were so unimpressed by not only English but European manufactures that they only wanted silver from the distant Westerners. They had in the relatively recent past built a fleet dwarfing anything Europeans could manage, while the Ottomans rampaged through the Balkans towards their heartland.

But say you give the aliens a hint and told them to look at Europe – who might they pick then?

Unlikely still they'd pick England, part of a soggy island off this distant peninsula, poor, undeveloped and regarded as barbarous even by its neighbours. London didn't compete with cities like Paris and Naples, which were three times bigger; nor was it as sophisticated as the northern Italian city states like Florence, Venice, Genoa and Milan. Only 3.5 per cent of the population lived in cities, compared to 10 per cent in the Low Countries. Nor was England particularly rich on per capita basis – poorer by a clear margin compared to Spain, Italy, France or the Low Countries. Its navy was weak, it had no overseas empire, its government bureaucracy was only a quarter as large as those of comparable nations, political and religious crises were routine, industry was almost non-existent, it lacked knowledge of sophisticated glassmaking or metallurgical techniques and needed imported German expertise to produce remotely up-to-date tools. Yet, three hundred years later, this small, damp country proclaimed itself master of the world at the Great Exhibition: richer, more powerful, technologically advanced and culturally influential than anywhere else on Earth.

You can read this two ways. First, that despite all the advantages of size and development it's still possible to throw away your advantage. England's 1550s Ming Dynasty contemporaries serve as another warning. But the more important lesson goes back to Chapter 1: that ideas themselves can catalyse extraordinary and unforeseen changes from even the most unpropitious starting points. If 1550s England could, despite everything, become a crucible for global takeoff then surely society today, with all our advantages, can do the same.

Post-coronavirus, a Great Acceleration could be starting to firmly counter the Stagnation.[13] Say we can buffer our institutions across the hurdles obstructing the next generation of tools. Moreover a huge effort is made to ensure they have a secure and 'provably beneficial' foundation. Suddenly other challenges become manageable. A new renaissance, an age of abundance and exponential improvement are no longer fantastical. Ideas we can't at present imagine revise our worldview, and something stunning emerges in its place. In the long term we're left with something like Iain M. Banks' Culture: benign, superintelligent AIs nurturing a humanity whose ailments are long forgotten.

Perhaps I'm just typical of our age, but betting the house on this happening feels a stretch. Nonetheless, such a conclusion is not, from today's starting point, absurd.

The more you look at these trajectories, one thing does become certain, however: the further the timescale is extended, the more likely either twilight or utopia becomes. From here the stakes only increase. Just the impact of something like an AGI or synthetic biology revolution or climate change or further pandemics sees to that. The middling confusion of recent decades cannot be sustained indefinitely. In the long term our ideas – and everything based upon them – will tend towards extreme outcomes; we will either rapidly shrink from the frontier or continue charging towards it. Ideas are so big and potent now that we are headed towards either conflagration or transcendence.

The Rocky Road

Zoom back into the timescale of the coming decades: barring a black swan event, we won't go in either direction. This in turn has two major implications: first, we will likely maintain a bumpy rate of ideas generation, and second, we can still influence the trajectory of the frontier's advance, nudging it one way or another, for better or for worse.

That makes right now a historical crossroads even more important than 368 CE or the 1550s. Now, as never before, the future of the planet and the species depends on our ideas. Every age believes it occupies the most acute axis of history. And yet, most of the time, history moves on, trend lines keep going. The arc is long. By now though the scale of the ideas are of a different order. With the nuclear bomb, humanity produced an extinction-level technology; it's plausible that we are on the verge of producing many more.

A macro history of ideas runs like this: general stasis, a gradual incline towards more ideas, a few periods of efflorescence followed by a sustained and gathering momentum across all dimensions, culminating in our own era of unprecedented production and blistering progress in some areas combined with surprising areas of struggle.

This will continue. That in turn means our choices over the next decade or two – both in terms of stimulating ideas, and also of channelling their direction – will ultimately shape the trajectory that leads to one of the two major outcomes. In the meantime we will follow our rocky road, as we have for the past fifty years, a road that will last until the present age is over – at the very least a few decades more. We will set our destination in the coming years.

But although there is uncertainty at the horizon, we also know where we are heading. This book's central thesis is that the West currently spends more to achieve less because it is doing the same things in the same way, over and over again, but that a global mix of ideas will break that stagnation to produce a new set of tools, which in turn can spark a wider acceleration in the production of big ideas. If you want to take one of the many strands explored here, this is, for me, the most significant for humanity's overall direction, and some variant of it will dictate the human frontier and where it goes.

For example, something like the following scenario could develop: society continues to converge at the frontier, adding wealth, education and capacity. However, shocks like coronavirus or endogenous financial crises flare up. Nonetheless, stirred by expanding new markets and international collaboration and rivalry, a generation of path-breaking, well-founded and properly designed breakthrough organisations, supported by a scaled-up world, become the locus of new ideas. These institutions, pockets of invention keeping the stagnant society at bay, build the new tools, decisively opening areas of enquiry or technology or imagination. Ultimately these tools outrun many aspects of the Idea Paradox. We will not climb the ladder unaided. At this point humanity enters the most perilous phase of the road – what form these tools and ideas take will dictate everything.

Over the medium term, R&D spend on the tools escalates. But despite the presence of those organisations, much of it is spent poorly and unevenly; focused on incremental innovations or rent-seeking behaviours or weaponised uses. Nonetheless, it's plausible to envisage huge ramps up coupled with tight focus – say a technological arms race between China and a Washington–Silicon Valley axis, ploughing the superpowers' best minds and resources into a bleeding-edge generative technology. As in periods of the twentieth century, meaningful competition spurs the United States to new heights. While AI might hit the conceptual buffers, it could lead to new vistas of quantum computing, spurred on by increasingly intense international rivalries and cooperations. Hopes for quantum computing then fall short. In the near term it has some impact on the cyber-sphere and on recondite areas of chemistry, but fails to break through.

Until, that is, an enterprising startup, a new IBM or even a consumer-focused company like Apple, finds a bunch of applications and a workable product, perhaps a biological application that helps a novel form of atomically precise fabrication.[14] Suddenly the world is changing. And, like the delayed impact of the Human Genome Project, over time the directions and possibilities of research are nudged forwards – biological or chemical computing is supercharged, and a new route to AI is found. At this point, a new UN-sponsored World Technological and Existential Risk Council is tentatively formed, pledged to try radical new forms of deliberation and collaboration, new cultural and political adaptations reshaping the global order in response, governments having learnt from the digital wave that farsighted management of technology is required.

The above scenario is indicative, but one could quickly think of innumerable others that fit the bill. Perhaps India, facing a failing monsoon and fighting an increasingly hot war with China on its Himalayan border, decides it needs to catch up on AI. Founding an All India Institute of AGI, it takes a radically open, collaborative approach to the task in contrast with China's secrecy. In concert with academic and industrial groups, this leads to a significant breakthrough: an open-sourced cloud-constructed Indian AGI.

Or perhaps the first $100 trillion company will come from somewhere multinationals have little impact – the slums of Rio, Lagos, Cairo, Addis Ababa, Kabul or Dhaka, where the physical and logistical infrastructure is inhospitable to existing Western players. Perhaps this could be a new form of business, a collective of quadcopter delivery drivers grown into a behemoth larger than the rest of the the Índice Bovespa or the Nigerian Stock Exchange put together, rewriting the rules of capitalism and collaboration even as it conquers physical-world challenges apparently beyond the reach of the tech giants. Or perhaps there are darker versions of all this, that reckon more heftily with the West's decline or geopolitical rivalry or extreme responses in the teeth of the climate crisis.

Equally any of those things might be spurs to action – climate change will be an immense challenge for sure, but meeting that challenge could prove transformational, just as the rise of China might re-energise the West.

One could spend days quite happily speculating, but the most important thing is the general space of operation: stagnation confronted with global scale-up, the confrontation of the paradox by a new set of tools. The horizons of the frontier are uncertain in detail, but solid in outline.

Five Suggestions

The human frontier is neither fixed nor a spectator sport. What happens next is down to us. So here are five broad suggestions for a flourishing future.

Go on a mission

Vannevar Bush had a good war. A former MIT Dean of Engineering and then the boss of the Carnegie Institution for Science in Washington DC, he saw early that the US had a problem: on the verge of the Second World War, American military tech and behind that its scientific base weren't up to scratch. More than for any previous conflict in history, this was a troubling disadvantage. Bush's idea was simple: gather smart scientists together, give them a lot of money, and task them with transforming that picture.

By the end of the war, thirty-thousand people worked for Bush, including two-thirds of all the physicists in the US. Funding had risen sevenfold on pre-war volumes.[15] But Bush wasn't done. With the war won, he saw the model's potential, as famously outlined in his 'Endless Frontier' report. He advocated the programme's ambitious expansion to win the peace: independent and well-funded scientists would build the future. Corporations couldn't and wouldn't fill the gap: only a massive axis of university-driven, federally funded research would work, underpinned by a new commitment to high skills training (epitomised by the GI Bill of 1944) to propel American leadership and advance. Despite resistance and scepticism, Bush, a skilled operator, swung the day. A programme of free science and military-led R&D began to roll out across the US. A further boost came with the Soviet launch of Sputnik, which galvanised US elites out of any sense of complacency and put a metaphorical rocket under the country's R&D capacity.

The results of Bush's vision were dramatic. In cash terms alone it's difficult to find a historical comparison. Between 1940 and 1964, driven by an animating purpose, American federal R&D funding increased twentyfold.[16] It was a river of 'endless money'. There were 130,000 people working on the Manhattan Project alone. The Apollo missions were five times as expensive, consuming 2.2 per cent of federal expenditure at their peak. A plethora of new bodies supported this work: 1958 alone saw the introduction of NASA and of DARPA, the Defense Advanced Research Projects Agency, responsible among other things for the Internet, as well as playing a part in the development of GPS, drones, stealth technology, flat screens and better artificial limbs. Just a few years on from NASA's creation, the US government was spending 0.7 per cent of total GDP on rocket and related technology alone.[17]

As a result, big idea after big idea was realised at scale. The period delivered an astonishing series of leaps in pharmaceuticals and medical treatments, from the rollout of antibiotics to the polio vaccine to the Pill; it built radar, the architecture of computing, modern telecoms and the Internet; it gave us rocketry, space travel, jet aeroplanes and satellites. More people than ever before had the chance of a university education and they enjoyed the fastest increases in median living standards the nation has ever seen. This was the era of Civil Rights and the welfare state, mass transportation and multimedia entertainment, universal education and healthcare.

Government funding and organisation, private enterprise and universities all came together to work on practical problems, technological innovation and basic research. None of these were seen as tradeoffs; rather they all cohered around the prospect of missions – getting to the Moon, building the bomb, creating a better world.

You might regard the moon landings as the most expensive symbolic gesture in history, but NASA missions spun off two thousand inventions and innovations including precision GPS, home insulation, baby food, ear thermometers, memory foam, wing and aerodynamic improvements, scratch and UV resistant glasses, air purifiers and the sensors used in digital cameras.[18] Satellites are now routine, from ocean rescue to weather forecasting. And of course, it all helped power the falling cost of computational capacity – computing's rapid take-off was underwritten by federal grants and the Department of Defense as the anchor customer in the postwar decades. At one stage the Apollo programme and its computing partner, the MIT Instrumentation Laboratory, was buying 60 per cent of all chips manufactured in America, an invaluable industry boost when demand elsewhere was only nascent. The lab's Software Engineering Division, under Margaret Hamilton, which developed on-board software for the missions, was pivotal in establishing the field.[19]

In the fifty years after 1951, this welter of innovation contributed to over half of America's economic growth.[20] Research demonstrates the value of Office of Scientific Research and Development (OSRD) contracts in the war (there were over 2000 of them, producing 7000 inventions,

2700 patents and 2500 papers): the epicentres of postwar industrial growth and R&D had received such grants.[21] So significant was this investment that the same clusters continue to shape the innovation and economic geography of the US – and the world – to this day. These investments mattered and were persistent, as Vannevar Bush had argued they would be.

That era ended. Federal research funding has been reduced to 0.7 per cent of GDP in total – $240 billion of funding is missing relative to its peak, down even from the early twenty-first century. The networks of government, business, universities, scientists, technologists and product people grew atomised and antagonistic. In particular the sense of mission, of a nationwide project to which all could subscribe, fell apart. It became an article of faith among governments that grand investments were a bad idea. The buccaneering world of missions and moonshots began to feel like a twentieth-century relic. Even organisations like ARPA took fewer risks as the intellectual confidence to back things at scale withered.

Big ideas need a mission renaissance. We need not just the expansive vision of a latter-day Bush, but the sense of urgency that compelled the epic scale of the Manhattan Project and the Apollo missions. Tinkering and serendipitous discovery long since stopped being sufficient. What worked for the seed drill or the steam engine does not work for nuclear fusion. In this new phase of our intellectual and technological history, something larger and more concerted is required.

This doesn't necessarily mean throwing everything at one programme in some monolithic recreation of the moonshot; but it does mean coordinating to mobilise large-scale research. Pierre Azoulay and Danielle Li explain how missions could help advance challenging ventures: a $1 million gamble with a 99.999 per cent failure rate isn't a good bet, but 200,000 such bets, a $200 billion research programme, would have an

87 per cent chance of success.[22] The only way you could ever fund those 200,000 projects, however, is by looping them into a clear and urgent goal underwritten with massive government support. Funders and researchers would otherwise, perhaps understandably, be too risk-averse.

Mariana Mazzucato advocates for such a mission-driven approach to grand challenges, harnessing the best of both public and private sector – as was the case in the postwar boom. Good missions, she argues, should inspire people, they must be bold, innately risky in their ambition, with clearly defined targets and deadlines.[23] They should span disciplines and sectors over the long term. Her approach doesn't imply a directed, top-down process – good missions are amenable to multiple approaches and experiments. They create and shape rather than negate markets; they should support and encourage bottom-up exploration. This is about 'picking the willing, not the winner'. There is no shortage of potential such missions, big ideas waiting to happen: mass-scale carbon-negative technology or fusion power or the elimination of dementia (or even ageing) or the creation of a permanent settlement in space.

Recent missions, in cancer for example, have lacked the urgency and commitment of Manhattan and Apollo. While there are positive signs, from the UN's rallying Sustainable Development Goals to funding boosts like the $100 billion Endless Frontier Act, and private initiatives from the likes of SpaceX (Mars) or Google (organising the world's information) to the massive spur in coordinated science in the wake of Covid-19, this weight of mission is still nowhere adequate. If even 0.5 per cent of global GDP was shunted into R&D, that would represent a $700 billion per year boost. Imagine if that was targeted to address our most severe challenges and realise the biggest goals.[24] Imagine what a rejuvenated, positive-sum

US–China special relationship built around concerted missions would do for the human frontier: surely it would be the biggest opportunity of our lifetimes.

A substantial body of research shows moreover that public money crowds-in funding, rather than crowding it out. Over 40 per cent of NIH grants are later cited in a private sector patent. Even Sergey Brin was at Stanford on an NSF dissertation fellowship. Military R&D alone has a big return – between $2.50 and $5.90 is returned for every dollar of Department of Defense R&D invested.[25] Increases in government-funded R&D are associated with major productivity improvements throughout the economy, and we saw how investments in the Second World War were still paying off decades later. Recall the spillover innovations of the NASA programmes; contemporary missions would bring their own. Small numbers of failures are used by political opponents to obscure vast amounts of good. All R&D is risky and being tolerant of failure is essential; the total return is massive.

Not only would ambitious missions be our best way of addressing challenges, they'd deliver a whole lot more on the side. If we want big ideas, we'll need to go on a mission to find them. We should.

Let a thousand experiments bloom

Remember when tech companies were still shiny corporate darlings? Back then Google was famous for a seemingly radical experiment in company management: 20 per cent time. Employees could use one day of their week to work on anything they wanted, experiment and try out 'wacky' ideas. It was so fundamental to company culture that it was referenced in the IPO. 'We encourage our employees, in addition to their regular projects, to spend 20 per cent of their time working on what they think will most benefit Google,' wrote the founders. 'This empowers them to be more creative and innovative. Many of our significant advances have happened in this manner.' And these projects *were* significant. Gmail, used by over 1.8 billion people, was one, as were Google News and aspects of AdSense, the money-making engine of the whole business.

It's been some years since anyone heard of a 20 per cent time project, or even of anyone at Google working on 20 per cent time. This is after all now a company that employs over 130,000 people. Engineers or creatives form an ever-shrinking core: sales people and lawyers and HR professionals dominate. What does an ad sales manager at a regional office do with 20 per cent time? Google also has more formal channels for R&D. It doesn't need thousands of people tinkering with fragmented projects and wasting scarce and expensive time. Officially 20 per cent is, one presumes, still there, but unofficially this bold move appears to have withered on the vine.

Which is a shame. Imagine if everyone in the world had 20 per cent time as part of their jobs. There would, for sure, be a lot of wasted time. But that freedom would produce results it's hard to quantify or envisage. We don't see how the steamroller of the everyday stifles ideas in offices, labs and studios. But I think it's safe to assume that, if 20 per cent time were not a strange outlier but a common element of many professions, we'd get a lot more interesting ideas.

How we go about having ideas is stuck. The structures and channels are broken but seemingly immoveable. We need to reimagine them. We need new ideas about ideas.[26]

There's no shortage of creative suggestions: abandon or transform peer review; distribute funding for projects randomly and blind – a lottery approach bypassing the issues of peer review which was recently tried with some success in Germany by the Alexander von Humboldt and Volkswagen Foundations; open funding processes to the entire population and let people vote; remove any human decision-making from the allocation process and leave it to a group of algorithms, tweak and experiment with them; have scientific teams crowd-fund their research.[27] Academia could trial career structures away from the 'publish in this journal, get a postdoc' path to accommodate more winding and eccentric routes.

We should experiment with an ambitious overhaul of the incentives structure in research. Earlier we saw how much exploratory and high-risk work goes missing thanks to the obsession with garnering citations. So let's change them. Jay Bhattacharya and Mikko Packalen argue that we should introduce a new metric which rewards research on the basis of novelty.[28] New ideas create new words and new word sequences; hence it would be possible to create a robust system that rewarded exploratory and high-risk research. This is just one suggestion in a burgeoning field of 'alt metrics' looking for better ways of incentivising and measuring ideas.

Perhaps grants could be awarded on the basis of how much help a team offers others around the world. Young people could be given less training and more responsibility; government R&D grants offered according to how much a company shares in the public domain. In the arts, funding often works to support non-commercial work; maybe that could be flipped so that government funding is provided for successful projects and people. Schemes could be introduced to discount credentialism and help, say, with late-stage career changes. Innovative funding mechanisms like Kickstarter could be further developed, new tools created for experiment. What should apply more widely to the domain of ideas: massive online collaborations like Wikipedia, techniques like red teaming and wargaming, novel approaches to computer simulations, randomised controlled trials (RCTs) in unfamiliar areas? A thousand experiments wait. Let them run.

A new generation of enlightened twenty-first century patronage could help – and not one which simply echoes the pathologies of the research ecosystem.[29] Patronage funded Descartes, Kepler, Galileo and Spinoza; and today patronage of a scaled-up modern variety could cut through the thickets surrounding creative thought. Alexey Guzey argues that there should be two elements to this. First, normalise long-horizon grants – award them for ten or twenty years. Then figure out the right environments for super-talented people to work in and double down on those. Rebuild Bell Labs – but for everything.

This approach is no hunch. One well-known study compared grants made by the elite Howard Hughes Medical Institute with those from the National Institute of Health.[30] Howard Hughes is high-minded, wealthy and private. NIH has plenty of cash but it spends public money and has to be accountable. Grants from the HHMI were awarded for a longer period, open ended in direction, tolerant of failure and encouraging of intellectual experiment. They lasted five years, usually, and often came with follow-on funding. NIH grants by contrast came with tightly defined deliverables, were one-offs and 'unforgiving about direction'. No surprises then that the researchers found breakthroughs were much more common under the HHMI scheme. The freer research was more creative, its impact greater. If you want radical experiment, then design for it; we don't.

Time and again in interviews for this book I heard the same thing: fund people and not projects, do so over long periods.[31] Remove short-term pressure for results. Create better, more balanced incentives. Find them by testing what works.

Prizes, like grants and patronage, can be a powerful driver of experimental work, whether in the arts or the sciences. But again, there is scope to open things up. Prizes drive people into the same old disciplinary niches. They reinforce boundaries. Why not design prizes to break those boundaries? Given the difficult realities of interdisciplinary work – which despite hype faces funding hurdles – we could do so much more.

Let's rethink the mechanisms of prestige in the face of a new, more complex reality. Large areas fall between the cracks of the Nobel Prizes' disciplinary boundaries (in areas that cross life sciences and physics for example, but also in AI research, network science and geology). Prizes freeze recognition in a nineteenth-century disciplinary straitjacket. What made sense to the Nobel Prize Foundation in 1895 doesn't make sense for the challenges and opportunities for research in the twenty-first century. Perhaps there should be a Nobel Prize for interdisciplinarity,[32] a Fields Medal for teamwork, a Grammy for breaking new ground. There are glimmerings of this in, for example, the Breakthrough prize or the X prizes. If we did the same in the arts, pushing beyond prizes that reward work in single genres or art forms, we could reimagine the whole economy of prestige, from the Oscars to the Booker. And the results would be spectacular.

Taking a more open and experimental approach to meta-ideas means trying more things – many more. It means creating regulatory space for risky experiments, not crushing them by committee or through an excess of precaution or the weight of vested interests. It means ignoring or bypassing inhibitory structures. There is no contradiction between missions and experiments; we need both. Good missions should enable rather than restrict experiments like these. Picking a side is more about the maintenance of a prior political vision than the prospect of encouraging and realising ideas.

The double Nobel laureate Linus Pauling once said, 'If you want to have good ideas you must have many ideas. Most of them will be wrong, and what you have to learn is which ones to throw away.' That's good advice. We need a set of experiments across the entire field of how we produce ideas. The sooner we start the better.

Relearn education

When discussing this book for research purposes, I'd always ask people what we needed to change. Increasing funding, producing a new kind of mission, was a response I expected. What I hadn't anticipated was the number who immediately said: education.

It has been an insistent hymn for a surprisingly broad spectrum of interlocutors. They argued that despite improvements we need profound global reform of education to meet the challenges of the twenty-first century; that many of the most productive areas for future research, in the margins of disciplines, powered by creative thinking, are still overlooked and underrated compared to the ever-present push for good exam results. We know this makes a difference – research already ties education styles and levels to rates of invention.[33]

There is a caricature of the educational process that it still embodies a quasi-military system: serried rows of rote-learning students facing the blackboard and diligently scribbling down notes from the almighty teacher. This is no longer true in many countries.

However there are still reasons to push for change. Education cannot adapt and evolve as quickly as other areas – for good reason, you can't always freely experiment with the futures of large groups of children or students, although there is still more scope for RCTs and iterative improvements. Then there is the charge, more difficult to dodge, that learning is excessively instrumental. Teachers teach to the exam, under pressure from governments and parents alike. Exams can be gamed, and they reinforce an increasingly dated parcelling of the world into tired subject categories. At high school and university, education has gone from a dull display of rote learning to a glorified box-ticking exercise. While governments everywhere profess education to be a major policy priority, it is often reduced, even at elite universities, to a blunt economic function: education churns out the next generation of productive worker-bees who can successfully manage spreadsheets. Hardly conducive to breakthrough ideas. People struggle to apply what they learn in a classroom, and it is even argued, given how much education is quickly forgotten and discarded, that it's an exercise in signalling (look how diligent I am!) rather than in gaining knowledge or building skills.[34]

It's striking how much education gets backwards. I am fascinated, even obsessed with topics relating to the study of physics or the origins of the first Industrial Revolution. I was taught both at school, and yet I, like most of my classroom peers, found them intensely uninspiring. How absurd! Some of this might be explained by the old adage that education is wasted on the young ... But of course it's also a matter of presentation. In the words of Matt Ridley, 'Science was taught – not just to me, but to my children as well – as if it was a catalogue of facts to be regurgitated, rather than a procession of fascinating mysteries to be challenged. Give them galaxies and black holes, not Boyle's Law!'[35] Quite.

At Harvard, E.O. Wilson's lectures always started with the broadest, most interesting questions and then worked back to the technical details. I was always amazed at university that the undergraduate physicists barely got to study the stuff that to me, an arts student, it was supposedly all about. It's easy to be a critic here, but the structure of most education simply isn't working as well as it could.

Much education isn't very effective even on its own terms.

Variation in quality within and between countries is large, but needn't be so. Many graduates with advanced degrees leave without understanding basic concepts in maths or science: in the UK, even (or especially?) the MPs governing the country get basic statistical and probabilistic questions wrong.[36] The Royal Society estimates that the UK produces only a third of the number of mathematically trained people it really needs. Compounding the burden-of-knowledge effect, two Yale scholars have documented an 'education-innovation gap' based on the analysis of 3 million syllabuses: many universities are teaching very far from the frontier.[37] Thanks to our poor short-term working memory, only about 30 per cent, or less, of a traditional lecture course really goes in. What's more, students often fail to see the big picture, their knowledge becoming more atomised and disjointed over a lecture series. Those in education often talk about the 4Cs of a twenty-first century education (critical thinking, creativity, collaboration, communication), but there is surely much further to go.

So let's look again. Perhaps we could redesign curricula around discovery and experiment; move away from ticking boxes, and towards imagination and the free play of ideas. Look at the successes of Finnish education, now an exemplar, and its principles based on teaching things like transversal thinking. Schools could aim to incorporate more insights from Montessori Schools, which educated entrepreneurs including Jeff Bezos, Larry Page, Sergey Brin and Jimmy Wales.[38] Their success hints at the potential in self- and peer-directed learning. In India a research programme showed how effectively school children learned on their own, unaided, when using and programming a computer. But the power of peer learning applies at university level as well.

The Nobel laureate Carl Weiman was shocked to find that after seventeen years of education in which they had excelled, his graduate students were still clueless about how to run a research project.[39] What happens next? they asked, stumped without hand holding. Worse, they didn't even really understand what physics was about, despite having degrees in the subject. Something was going wrong. So Wieman created the Science Education Initiative (SEI). Instead of standing and delivering facts, Wieman advocates a science education that helps students learn and explore for themselves. It facilitates independent thought, allowing students to proactively piece together the puzzle. Peer instruction is key: students who help teach each other are more engaged, tripling their knowledge gains over a course. Students understand how their peers think, and tailor their instruction accordingly. Psychologically, they are ready to engage with one another in a different manner to the formal teacher.

Such ideas are only the beginning. Teachers and researchers want to isolate and roll out best practice, whether for four-year-olds or high-achieving undergrads. Some improvements are quite basic: if teachers slow down delivery, it helps learning. Tutoring, and within that methods like direct instruction and mastery learning, can help anyone learn better.[40] Technology is still only at the beginning of its pedagogical journey, albeit greatly spurred by the lockdowns which were a consequence of coronavirus: simulations, long-term assignments, radically personalised learning and curriculums could all make a difference. Making software work effectively for education is tricky, but again, evidence suggests it can be done.[41] Universities like MIT, Tsinghua, Imperial College and the Oxford Martin School are starting to replace traditional courses with a more problem- or topic-driven approach, cutting across traditional subject domains to address real, transdisciplinary issues. We should roll out a new era of leaner, faster, more application-focused PhDs.

Change in education is always hard. Wieman's efforts are

a cautionary tale: once the grant money ran out, momentum stalled. Let's make sure that doesn't happen on a generational scale. Let's relearn education.

The other institution revolution

In colonial Delhi, the authorities were concerned about a rash of deadly cobra. What to do? Offering a financial reward for anyone who could produce a dead cobra seemed sensible – outsource the solution and get everyone cobra hunting. That'll sort those cobras! Only the upshot was rampant breeding as people aimed to clean up on the reward. Similar incidents are alleged to have taken place in Hanoi with rats, and under Chairman Mao, whose Four Pests Campaign wiped out the sparrow population, leading to a huge increase in the number of insects and ultimately, in part, to the Great Famine. These stories are about perverse incentives, a particularly piquant example of a timeless problem: that of unintended consequences.

The history of ideas is littered with unintended consequences: all those serendipitous moments and accidents, for a start. The great theorist of serendipitous discovery, Robert Merton, also wrote a study of unintended consequences exploring how the pressures of time and money, the complexity of the world, the capacity for error, all promote the free play of unintended consequences, rippling out in multi-layered and unceasing chains.[42]

As we saw, there is nothing uncomplicatedly 'good' about big ideas. Weapons and modes of war are ideas just as much as medical treatments or great works of literature. Ideas can have a devastating impact: think of the consequences of naval and navigation technology for the indigenous peoples of America, or the roiling conflicts seeded by the Reformation. Fertilisers fed the world and combustion engines got it moving; neither was meant to damage nature, but both did. No one can work through all the manifold consequences of a thought or action, especially where it is inherently novel. The world's complexity means understanding even first order, let alone second, third or fourth order consequences is an immense challenge. Even then, the intention to produce only good outcomes from the acceleration of ideas is potentially problematic.

What now adds urgency is the existential nature of our ideas, from the Bomb to the melting Arctic ice. At one level this is systemic: about a 'risk society' where almost everything from CFCs to social media has the potential to cause collateral damage.[43] The Internet used to be exciting because of the creative collaborations, citizen journalism and networked science it enabled. Instead it first became about time wasting, pointless rancour and advertising, and then about an exercise in monopoly capitalism and precision-guided propaganda. With technologies like AI and synthetic biology and the scale and power of super-industrialised civilisations, big ideas now have species-level consequence. This time really is different.

Such cascading impact is also found in the social sciences. It's a curious irony that while Karl Marx dismissed the 'Great Man' theory of history, he personally altered the trajectories of entire nation states. The mind boggles to think how, in a little under a century, the mind of one mid-Victorian scribbling in the Reading Room of the British Museum impacted billions of lives, with reverberations right up to the present. The small group known as the Mont Pelerin Society, a collection of academics, policy wonks and business executives, managed first to sketch then to establish the neoliberal framework that has dominated international society since the early 1980s. Suffice to say that each side has vociferous critics. The point is, the world is prey to ideas, whether or not they are wanted or beneficial or thought through; good and bad ideas alike bubble up and are foisted upon us.

Can we do better? I think so. That means producing a new set of institutions, norms and mechanisms for working with ideas; for helping them quickly progress in a world that blocks them, while at the same time mitigating the most serious downsides. Barriers from over-regulation and bureaucracy to corporate cronyism are already suffocating risky work. We're thus on the horns of a dilemma: on the one hand it would be reckless to simply let things happen; on the other, we already stifle too much.

This is the great institutional challenge of the twenty-first century. A challenge at least as hard as realising AGI or elaborating a Theory of Everything.

There will be several levels to any response. First is to build ethical and safety oversight into research and ideation. Both have become major topics in AI research, for example, which is encouraging, as if environmental factors were taken into account by James Watt or Karl Benz. But there is so much more to do, especially given that this focus is hardly uniform: some groups care deeply, others probably less so. Over the coming years we will see more forms of what the life sciences call 'dual-use research of concern' – essentially where, say, academic biosciences research could be repurposed to produce new biological weapons. Such work is flagged early and triggers more careful publication and dissemination protocols. Above this come questions of innovation-enabling regulation, an increasingly intense and exciting area of research where massive gains are still to be found.[44] Choosing the right kind of risk-minimisation strategies will be critical.

We *need* big ideas about how to do this. Our best thinkers will have to raise their game. Doing so could produce the signature concepts of the century. Without them it's not implausible that societies will clamp down on new tools, finding doors to discovery shut by fear; or, worse, will deliberately pursue those tools for malign ends. But there is cause for optimism – a new generation of ideas in philosophy, law, economics and social policy will arise from grappling with the radical implications of powerful, widely dispersed technology. Even the processes used to arrive at these ideas may be distinctive – for example, cross-border deliberative mechanisms capable of global mediation relating to uncertain but immensely powerful technologies. Before you can (try to) instil values in an AI, you first must agree on those values, and agreeing values that represent humanity is difficult to say the least. But thus it ever was at the frontier.

One area where low-hanging fruit can certainly be found is in the regulatory and policy environment. There are many potential wins – including for example a global rollout of R&D tax credits ('a 10 percent fall in the tax price of R&D results in a least a 10 percent increase in R&D in the long run'), or a radical overhaul of the IP regime, or a redesign of the structures in academia, or a review of how governments fund creative work.[45] We also need better forms of social accounting.[46] For a venture capitalist, the input and the output ultimately tally: it's about money in and money out. This doesn't work with regard to many ideas. Science, for example, requires enormous deployments of capital, but the payoff cannot be understood in purely monetary terms. The same is true of philanthropic efforts or the humanities. So if we are to encourage big ideas, and good ones, a fuller and more multi-dimensional social accounting of ideas of all kinds is required, one capable of more diverse metrics, seeing wider levels of contribution.

Lastly, there is scope for new or certainly better supranational institutions.[47] Ideas, from carbon-negative technology to massive multiplayer online games, are global. So are the unintended consequences. Just as we are entering a resurgent era of great-power competition, we need meaningful collaboration here. Anyone with an investment in new technologies, organisations, capacities and collaborations, which is to say all of us, can and should be thinking about this.

Get this right and it would be a huge spur to ideas, giving us much greater confidence to experiment. Far from inhibiting research, done correctly it would create the necessary societal space and institutional foundation. Get it right and we secure the future – and set the ground for the last recommendation.

Go bolder

It's notable how many scientists with major discoveries to their name are also musicians, writers, philosophers, activists and communicators. Physicists like Einstein, Richard Feynman and Werner Heisenberg were accomplished musicians; Marie Curie, Fritz Haber and Erwin Schrödinger wrote poetry; biological and medical researchers from Louis Pasteur and Joseph Lister to Alexander Fleming, Howard Florey and Dorothy Hodgkin were artists. Is there a connection?

Looking at hundreds of discoveries and significant scientists from the UK, France, Germany and the US over the twentieth century gave J. Rogers Hollingsworth a clear answer: those with what is known as 'high cognitive complexity' were much more likely to make major discoveries.[48] Scientists capable of internalising many different contexts, backgrounds and fields and hold them in their heads at once, comfortable with the ambiguity and contradiction, who were prepared to look from multiple viewpoints, who drank in complexity and range, had the biggest success.

The more specialised and comfortable researchers get, the more difficult this kind of innate cognitive diversity becomes. All that art and politics is indicative of the rich multiplicity of a mental world and it leads to qualitatively different styles of discovery. The best scientists, for example, zigzag and switch topics, changing with surprising frequency: 43 times in their first 100 papers.[49] Even as we are incentivised to burrow deep, it is essential to go wide, to swim out of our depth. This also applies at the level of the research project.[50] Yes: the more improbable and out there, the more likely a project is to be significant.

Big ideas are surprising. Researchers examining the US patent record and over 19 million biomedical articles dating back to 1865 found that surprise, in both the context and the content of the work, was important (e.g. surprise in context is publishing in a journal far away from previous research). The pattern is unambiguous and revealing.

Overall the probability that a paper or patent would prove to be of major importance increased with the novelty of content and context. Novel papers were five times more likely to be a 'hit'. Unusual backgrounds and collaborations were helpful: again, these surprising personal factors made a difference. However, even more important than either were 'knowledge expeditions', where researchers strayed far from their home territory into new disciplines and problems. Indeed, such 'expedition novelty' emerged as the strongest predictor of breakthrough discoveries. Scientists prepared to take leaps into faraway fields were the most impactful of all.

Separate analysis of 17.9 million scientific papers across a range of fields suggests the same result: while the most impactful papers have conventional linkages, they usually also have something radically 'atypical' within them.[51] Unusual knowledge and out-of-context information and techniques tend to make themselves felt. This shouldn't be news, inasmuch as the role the unexpected plays in science has been a major theme from Robert Hooke to General Electric labs.[52] But in an age of planning, caution and predefined expectation it gets lost. Expeditions far outside home turf are rarely encouraged.

There is a common thread here, and one that recurs again and again in the literature: big ideas form when people and teams step outside their comfort zones, when they bridge distant worlds, when they combine the seemingly disparate and see the patterns no one else can.

This requires an all-too-rare mentality. It is about what David Epstein calls 'range', leaping beyond and between sanctioned silos, about an embrace of change, breaking boundaries and rules, a militant attitude to complacency, about a vaunting, at times frightening, level of ambition. It is about being wide open to everything, however random, forcing yourself to see things anew. There is also a hefty dose of dogged resistance to small-world scepticism, groupthink, authority, accepted boundaries, standard practice and tradition. It's a refusal to accept that we are destined for slower, lower rates of progress and advance.

If I had to boil it all down and deliver the message of this book it would be:

Step outside the usual channels. Wander far and free. Go where it feels uncomfortable and uncertain, where the water is deep. Do not conform. Try things that are unlikely to work. Try again when they don't.

Take more risks.

In writing this book, I traversed a number of fields. Each is full of the smartest and most creative people in the world. But it is incredibly easy, almost essential, to become wedged deep in a disciplinary silo, stuck in a worldview. The best economists cannot help but think like economists. Artists think as artists – they care about the attention of their peers, they are caught up in the rituals and mores and institutions of artists and art teaching and dissemination. As disciplines' frontiers progress, so they move further from others, and as they do they curdle, turn inwards, beginning to self-converse and, lacking the energy of the unexpected and different, move forward only in self-referential increments. That needs resisting.

Above all, we shouldn't fear failure. That point is always one of those Silicon Valley rallying cries: fail fast and all that. In general I'm sceptical when people talk about embracing failure. It's usually the richest and most successful who tell the rest of us that failing is just fine – it's classic survivor bias. Meanwhile, back in the real world, failure sucks: we don't have endless chances, it hurts and destroys careers. Even in the most benign places failure hurts. But it's also essential.[53] Trial and error are not optional components in pushing back the frontier. For individuals, organisations and governments, making breakthroughs means being massively comfortable with failure.

If you're funding people, some will spend twenty years with nothing to show for it. The pioneers of mRNA vaccines spent decades in the academic and commercial wilderness: ignored by funders and journals, they were even, in some cases, demoted because of their 'wasted' research.[54] That is the price of progress. Innovation *is* risky. Lady Mary Wortley Montagu, who introduced the practice of smallpox inoculation to Britain, first tried it on her own son. The aviation pioneer Otto Lilienthal, who so inspired the Wrights, died in his flying machine. Tinker with a steam engine, and it might blow up in your face. Write the wrong article in the wrong pamphlet and go to prison – or worse.

I'm not saying experiment on your children or throw yourself off a building ... But if it doesn't feel close to that level of commitment, go bolder. Create a new breed of mission to tackle our most daunting challenges. Remorselessly experiment with how ideas are produced. Revolutionise education to produce hungry, roaming, courageous minds. Establish the right foundation for risks – and then take them. Perhaps the best thing that can be said for the current turbulence in the world, the pandemic and the uncertainty, is that it is, or should be, a prompt to ambitious rethinking. A time for rebirth, renewal, re-imagining, when the risks of stagnation are clear.

The human frontier is our responsibility. In the long term, it will not bounce along in an equilibrium or gradual incline. There will be no reversion to the historical mean. Its direction is still being set. We have to get this right.

Epilogue

Paradigm Unstuck

In the 1960s the Soviet astronomer Nikolai Kardashev was thinking about how we might detect extraterrestrial civilisations,[1] what signals they might send and how we should interpret them. He realised that while such beings might differ in almost every possible particular, they would still harness energy. He produced what became known as the Kardashev Scale, a ranking of civilisations' energy use, and a proxy for measuring the ideas frontier at the grandest level.

A Type 0 civilisation is like our own: able to access some reasonable portion of the energy falling on its host planet. A Type I civilisation, a truly planetary form, is able to access and store all of it. Above that come stellar civilisations, Type IIs, capable of using the force of an entire star. And then, much more advanced still, are Type III civilisations, entities that control the energy of a galaxy. Some years ago Carl Sagan put us at 0.7 civilisation. The physicist Michio Kaku puts us at 100 to 200 years away from becoming a Type I: the next key threshold for humanity.[2]

Let's assume a positive version of this book's central thesis, that humanity builds the next generation of institutions, scales up, finds new tools – that we follow the trajectory of the past centuries, that big ideas accelerate rather than diminish. Loosely adapting the Kardashev Scale helps us think about where we are headed. So let's have a guess – and some fun.

Type I Ideas (near term-ish)

New technologies emerge and drive change, from room-temperature superconductors to atomic transistors; the Great Stagnation is long over. Thanks to a surge of new break-through institutions, we achieve full quantum advantage, and machine learning is deployed with awesome power. Biological and chemical computers will redefine how we compute. AR and VR, with full haptic capability, finally reach critical mass and become everyday aspects of life, altering forever the day-to-day experience of reality. With AI, this nullifies the burden of knowledge effect. Fully personalised and remote medicine are available at marginal cost, at a stroke eliminating Eroom's Law and indeed most ailments: the greatest shift in medical capacity since Pasteur. Naturally this also means finding cures for cancer and dementia. Pandemic response is transformed by auto-vaccines. We reach Mars and establish a permanent settlement there, helped by huge advances in robotics, rock-etry and self-assembly systems. Technology also fights back against climate change: from vat-grown meat, to genetically modified plastic eaters, to direct air capture, to space-based solar power, there is, in the end, no shortage of answers.

Much else is revolutionised. This novel toolkit re-accelerates fundamental science, allowing researchers to unravel the dark matter and dark energy, profound mysteries that sit at the heart of our understanding of the cosmos. We might even detect the 'dark force', an even more mysterious fifth force of nature acting as a bridge to dark matter and energy. We should expect great and popular new art – the Shakespeare of computer games, the Beethoven of VR, the J.K. Rowling of DNA. There is scope for major new genres and new media to emerge. Perhaps we might one day have an installation artwork with viewing figures equivalent to the most popular TV programmes. We might, eventually, get the Great American Novel.

In philosophy we develop a new bioethics for the emerging possibility of transhumanism, even as the first posthumans develop a unique culture. Experimental philosophy flowers into a massive and rich discipline, as subject categories fixed for so long in schools and colleges and journals dissolve and blend into exciting, intriguing and productive new forms: a great reordering and cross-pollination of knowledge and practice. This might help render a complete theory of human behaviour and account of the origins of human culture.

Economics undergoes several further paradigm shifts, finally dropping its neoclassical template for something more sophisticated, perhaps working closely with advances in biology. New theories emerge about how to build truly resilient economies and societies, capable of withstanding and growing from unforeseen shocks. No more complacency. The processes of invention, innovation and collective intelligence are studied with even more intensity, yielding the nearest thing we can get to complete and normative descriptions that sector by sector retire the breakthrough problem. In the wake of the Great Convergence, growth is in orders of magnitude. Using the next generation of computers and revolutionary economic models, the emerging science of cliodynamics – the statistical prediction of society – grows apace. Eventually it is used like a weather forecast. As businesses and universities morph into unfamiliar forms, hundreds of millions now work in hybrid, self-sufficient and self-organised creative research collectives.

Charter cities, perhaps floating on the sea or in orbit, become hotbeds of political experimentation and usher in a new, more fragmented polity makeup. These freewheeling new statelets are the locus of much of this innovation, themselves representing the biggest experiment in statehood since the birth of the nation state.

As vast photovoltaic sails pitched around the Earth suck up more and more of the Sun's rays, we approach Type I status. After this, things get murkier.

Type II Ideas (medium-term ish)

A stellar civilisation finally sees the creation of a post-scarcity society, a new economic, social and political domain, with a wholesale rewriting of the legal and social-scientific framework around it. Old notions of the economy are redundant. Political ideas like left wing and right wing are ancient and irrelevant relics by now, and instead a series of new oppositions has arisen. A delicate system of transnational decision making handles all major questions – not least about the new range of technologies being created and the great pan-Earth missions embarked upon. At the heart of this system are new moral principles, post-ideological codes whose assemblage makes them akin to the creation of a post-religious moral order equivalent in scope to Christianity, Islam or Buddhism.

Scientific research takes another astonishing leap. The riddle of consciousness is solved, including the nature and experience of non-human intelligence. Things that now might be called paranormal are found to be grounded in reality. Indeed, the growing respectability of panpsychism, the idea that all matter has an element of consciousness, is experimentally validated. Speaking of which, SETI finally makes a breakthrough. Astrobiologists can even speculate what these distant entities look like. Proof of the origins of life on Earth is established. We don't quite find a Grand Unified Theory of Everything (that phrase being no longer in vogue), but an understanding of quantum gravity opens a more fundamental realm of physics altogether, doing to quantum mechanics what quantum mechanics did to Newtonian physics – and no, this isn't proof that string theory is right!

All of this underwrites further evolution of technology. Fusion power is everywhere, and in miniaturised form. The Mars settlement leads in terraforming the Red Planet, even as an asteroid mining station on Ceres helps establish a settlement on Titan. So-called fifth wave technologies begin to emerge: antimatter engines, light sails, ramjet fusion engines and nanoships. Nanotechnology assemblers are a reality, rendering the physical world an effortlessly malleable creative platform, even as robotic fabricators grow to the next level of sophistication, constructing vast gossamer megastructures in space. Yet more significant are the first fully functioning Von Neumann probes: self-replicating robots thrown out into the galaxy to endlessly reproduce themselves, ultimately to form a numberless swarm strung out across space. A full Type II civilisation is eventually realised with the construction of a Dyson sphere around the Sun, collecting all its energy.

Closer to home, whole-brain emulation is a reality, ushering in a new em civilisation in parallel. Everything changes again when artificial general intelligence finally becomes a reality and the intelligence explosion kicks in.

Type III Ideas (long-term)

At this distance, ideas grow hazy – like gazing at a distant star, their light fades in and out, blurry and indistinct.

Superintelligences master faster-than-light interstellar

travel. Terraforming becomes near-automatic thanks to a 'Genesis device', the seed for creating habitable planets. The great panoply of exoplanets becomes the playground for our descendants. It's difficult to guess at the creative or political imperatives here, but these ideas are on a grand scale: art works the size of solar systems or even galaxies. Energy is harnessed on an epic scale; perhaps dark energy is used, perhaps we can create negative energy or access Planck energy.

Mind-bending ideas are the stuff of the everyday. All knowledge is unified. Religious and metaphysical questions fuse with what was once called science; the theological and the material are, at last, a continuum. We find answers to questions of the multiverse – were there many big bangs, infinities of prior and parallel universes? Do they all have different laws of physics? How might those other laws operate? Can we travel between them and make use of them? We might even definitively prove the universe is but a simulation running on some distant server – and, having deduced this, learn how to crack the code of the simulation and rewrite the laws of nature themselves. Only here are the limits of the Idea Paradox truly breached. Speculation yes, and speculation necessarily – ideas have now gone beyond our modest and squishy brains.

Back in the realms of Type 0.7 civilisation, this is some of what we could expect at the frontier, near and far, in future – and they are big ideas indeed, far beyond a 'human' frontier. If this road is taken, then the ultimate fate of the human frontier is to render itself an historical artefact rather than a living threshold.

Perhaps the very fact we can briefly sketch an outline like this is another reason for optimism: the adventure of ideas and the great project of the frontier is far from over. And that highlights something else. If what we can see from a twenty-first century 0.7 perspective exhausts the possible, that would be unusual. We should be excited most of all by our blind spots, the ideas we can't or don't imagine; the portions of thought so radical and strange that discussion is still quite literally inconceivable. The biggest ideas of all will be those that truly surprise us.

As long as ideas can do that, the future is bright. The frontier awaits.

Notes

PROLOGUE

1 Di Pasquale (2018), p. 14
2 Ibid., p. 29

INTRODUCTION

1 These frontiers are clearly very different in character. The knowledge frontier can be agreed on more readily than the aesthetic frontier. Yet in the latter case, new media and stylistic forms can be seen as broadening the possibilities of experience in an analogous way, even if we should be alive to the various characters and imperatives of the many different frontiers.
2 Khan and Wiener (1967)
3 Greenspan and Wooldridge (2018), p. 35
4 Ibid.
5 Graeber (2016), p. 105
6 Douthat (2020)

1 HOW BIG IDEAS WORK

1 Mokyr (2018)
2 McCloskey (2017)
3 Ibid., p. xiii
4 See both Mokyr (2011; 2017), and Henrich (2020) for the deeper underlying psychological component behind this.
5 Howes (2017)
6 Jacobs (2014)

7 Mokyr (2017)
8 Many scholars come down on one side or another, but as the entire oeuvre of Joel Mokyr goes to show it's almost always more complicated than that.
9 I would like to thank DeepMind Research Engineer Daniel Slater in particular for his suggestions with regard to this phrasing.
10 See for example Rogers Hollingsworth (2007) for this model of ideas.
11 Kuhn (2012), p. xliii
12 Ibid., p. 174
13 Ibid., p. 85
14 Ibid., p. 207
15 Gould (2007), p. 3
16 Rather than a straight porting of the concept he was willing to explore 'homologies', but only in a limited set of circumstances. Ibid., p. 266
17 Mokyr (1999)
18 Christensen (2013); Perez (2002). On a more local level we might look to what Louis Galambos called 'formative' innovations and Clayton Christensen 'disruptive' innovations.
19 Bahcall (2019)
20 Thiel (2014)
21 Kelly et al (2020)
22 For an alternative methodology with technology see for example Packalen and Bhattacharya (2012) or Funk and Owens-Smith (2017).
23 See for example a paper like Bhattacharya and Packalen (2020) for means of categorising science.
24 See Jockers (2013) or Moretti (2013)
25 Henrich (2009)
26 Epstein (2019)
27 See Hargadon (2003) for the concept of 'technology brokering' more widely, or Ridley (2020) for more on recombinant innovation. See scholars of creativity like Howard Gardner for more on the centrality of recombination to creativity.
28 Koestler (1970)
29 For much more detail on this see for example Johnson (2011), pp. 28–9; Koestler (1970), p. 119; Henrich (2009).
30 Le Fanu (2017a); Henrich (2009)

31 See Merton and Barber (2003) for many instances of the importance of serendipity in science, technology and even the humanities and politics.
32 See for example Johnson (2011) or Henrich (2009).
33 Merton (1961)
34 Ridley (2020), p. 119
35 Michaël Bikard (2020) has created a database of 'idea twins' with some 10,927 entries. For more general background see Gladwell (2008).
36 Diamond (2005)
37 Tuchman (1985)
38 Mokyr (2014)
39 Tetlock and Gardner (2015)
40 Kay and King (2020)

2 THE BREAKTHROUGH PROBLEM

1 Koestler (1970), p. 112
2 Inoculation, the principle behind vaccination, had been known about much longer.
3 Smith (2012)
4 Le Fanu (2011), p. 14
5 O'Mahony (2019), p. 7
6 Polio is still to be found in Afghanistan and Pakistan, but elsewhere has been fully eradicated.
7 Le Fanu (2011), p. 353
8 Gordon (2016), p. 124
9 Ibid., p. 216
10 Ibid., p. 228
11 Ibid., p. 209
12 Office of National Statistics (2015)
13 Ibid.
14 The numbers have been widely reported and discussed. See for example Roser (2020).
15 See for example Therrien (2018).
16 Scannell et al. (2012)
17 Ibid.
18 Erixon and Weigel (2016), p. 142
19 Le Fanu 2011, p. 283

20 Ibid.
21 O'Mahony (2019), p. 11
22 Conversation with Jack Scannell, 5 April 2019
23 Le Fanu (2011), p. 6
24 Kings Fund (2020)
25 Scannell et al. (2012)
26 Pinker (2019), p. 333
27 Cancer Research UK (2019)
28 O'Mahony (2019), p. 145
29 C. Graeber (2018), p. 5
30 Account drawn from McCullough (2015) and Weightman (2015); see either for more details on the Wright brothers.
31 Greenspan and Wooldridge (2018), p. 53
32 Ibid., p. 55
33 Ibid., p. 96
34 Smil (2020), pp. 181–4
35 Mokyr (1999)
36 Indeed, an innovation essential to the Wrights.
37 Greenspan and Wooldridge (2018), p. 263
38 See Gordon (2016), Chapter 11 for more details.
39 Coggan (2020), p. 328
40 Smil (2020), p. 183
41 Although arguably not better in all respects. The weight-to-payload ratio of modern cars is actually worse than for older vehicles.
42 Other areas which may also have a breakthrough problem of a similar structure include energy, agriculture, construction and manufacturing – and also areas like science, culture and business growth which we will turn to.

3 THE DIMINISHING REVOLUTION

1 Mahon (2004), p. 163. The following account of Maxwell is derived primarily from the same source.
2 Ibid., p. 65
3 Huebner (2005)
4 Developments as outlined in *The History of Science and Technology* by Bryan Bunch and Alexander Hellemans.
5 Huebner (2005)

6 Vijg (2011), p. 28. See also https://en.wikipedia.org/wiki/Timeline_of_historic_inventions
7 Nor is Vijg's the only other taxonomy to show this. See Cowen and Southwood (2019) for a more comprehensive survey which suggests that wherever innovation is tracked over the long term something similar emerges.
8 Smil (2005)
9 Ibid and Mokyr (1999)
10 Greenspan and Wooldridge (2018), pp. 132–3
11 Ibid.
12 Smil (2020) and (2005)
13 Mokyr (1999). It is worth noting that Mokyr is much more of an optimist about the future of big ideas than other thinkers discussed here.
14 Smil (2005)
15 See for example Gordon (2016); Cowen and Southwood (2019)
16 Vollrath (2020)
17 Gordon (2016), p. 2; Vollrath (2020b) In Cowen and Southwood (2019) there is a developed discussion that deals with many of the conceptual criticisms of the TFP as a central plank of evidence in the stagnation debate.
18 Cowen and Southwood (2019)
19 Ibid.
20 Vollrath (2020a) in particular makes this case strongly. However his thesis is not inconsistent with the stagnation effect also being strongly apparent.
21 Kelly et al. (2020)
22 Gordon (2016), p. 566
23 Wolf (2019)
24 Quoted in Simonite (2014). See also Scheu (2019) for more of Thiel's thoughts on innovation and stagnation.
25 Although this is widely argued, I am indebted to Benjamin F. Jones for forcing home the power of this line of argument.
26 Discussed in Bannerjee and Duflo (2019), p. 187. See also an extended discussion in Cowen and Southwood (2019) of why digital does make such a difference.
27 See for example Colvile (2017) or innumerable op-ed pieces casually making the same claim.
28 For the impact of energy on civilisations, see Smil (2017).

29 Storrs Hall (2018)

30 Of course, while there are signs of an energy transition to renewables at the frontier as well, this is not contra the Henry Adams curve delivering a thirtyfold increase in available energy. See Storrs Hall (2018) for more detail on why the curve flatlined and how it could change in future.

31 Dorling (2020)

32 This wording – an age of consolidations not revolutions – comes from a conversation with Peter Watson, while the quote is from Patrick Collison in Smith (2021).

33 There is a considerable debate on this question. Some argue that the gains are still delayed and yet to show up. Others that much is missed in GDP figures, for example, because so much is intangible. While the latter in particular has force, surely the most telling thing is that the debate is needed; it would have been much harder to claim that the fruits of the 2IR were not revolutionising all of society.

34 Warsh (2007)

35 Ibid., p. xxii

36 Yueh (2018), p. 266

37 Solow (1956)

38 The great economist Kenneth Arrow, among others in a long intellectual lineage, can also claim credit for establishing the link between knowledge and increasing returns.

39 Romer (1990)

40 Greenspan and Wooldridge (2018), p. 361

41 Jones (2019)

42 Jones (1995)

43 Ibid.

44 See also Laincz and Peretto (2006) for a similar analysis.

45 Jones (2002); Cowan (2011), p. 18. The literature and models for growth theory are clearly much more complex and developed than those outlined here. See for example: https://bcec.edu.au/publications/the-decades-long-dispute-over-scale-effects-in-the-theory-of-economic-growth/

46 Bloom et al. (2020)

47 Ibid. All figures below taken from this paper unless stated otherwise.

48 Ibid.

49 See for example Wong (2017)

50 Bloom et al. (2020)
51 Cowen (2011)
52 Ibid., pp. 14–15. These figures are calculated by Cowen in 2004 dollars.
53 Thanks to Mikko Packalen for this and many other suggestions.
54 See for example Erixon and Weigel (2016). It could be argued that European innovation is concentrated in smaller companies like those forming the German Mittelstand. However that then prompts the question of why Europe cannot scale innovative companies – another but related problem in executing big ideas.
55 Erixon and Weigel (2016), p. 11
56 Greenspan and Wooldridge (2018), p. 395
57 Naudé (2019)
58 Cowen (2018a), p. 6
59 Ibid., p. 73., see also Shambaugh et al. (2018). The point isn't universally accepted, however, and many would argue we are seeing record tech company creation.
60 See for example Madrigal (2020)
61 Decker et al. (2014)
62 Shambaugh et al. (2018). In fact, they have tended to increase their market share over that time.
63 Erixon and Weigel (2016), p. 192
64 Vollrath (2020), p. 144
65 Akcigit and Ates (2019)
66 Greenspan and Wooldridge (2018), p. 396
67 Lafond and Kim (2017)

4 THE ART AND SCIENCE OF EVERYTHING

1 Quoted in Moore (2019), p. 15
2 Ibid., p. 107
3 Ibid., p. 121
4 Quoted in Illies (2013), p. 121. Information on 1913 largely drawn from Illies.
5 Ibid., p. 247
6 Ingham (2019)
7 Andersen (2011)
8 See Douthat (2020), p. 112 for his list of distinctions of the vital and decadent culture from which this springs.

9 Campbell and Mathurin (2020)
10 Kim (2017)
11 Fernández-Armesto (2019), p. 331
12 Ibid., p. 341
13 Watson (2001)
14 See for example Sand (2018)
15 Graeber (2016), p. 134. He is not alone in thinking this, especially about social and cultural theory. 'The golden age of cultural theory is long past' wrote Terry Eagleton – in 2003 (Eagleton 2003). Meanwhile the social sciences exhibit, in the words of one economist, 'slow progress', with few if any recent breakthroughs in 'psychology, sociology, economics, and political science' (Clancy 2020b).
16 Callard (2020)
17 Burke (2020), p. 246. Those it lists as being born in the 1950s include Judith Butler, Daniel Levitin and Robert Sapolsky. The numbers are poor compared to preceding decades and no better than preceding centuries.
18 Fukuyama (1992). Psychologically satisfying is here meant in the technical sense Fukuyama meant it: liberal democracies enable recognition, a core human desire, in ways other forms of political organisation do not.
19 Collier (2018)
20 Other candidates might include Modern Monetary Theory, or Charter Cities.
21 See for example the work of Geoff Mulgan on this and the absence of ideas in the arena of the 'social imagination'.
22 Some may of course question whether this is a good thing; but regardless, it reflected a novel agenda and ambition realised at scale.
23 Critiques of science of the kind presented here are not necessarily recent – see for example Platt (1964).
24 Sarewitz (2016)
25 See for example Wilson (2017), p. 192
26 Landhuis (2016)
27 Ibid.
28 Gusenbauer (2019)
29 Ibid.
30 Arbesman (2012)

31 See Fortunato et al. (2018)
32 Collison and Nielsen (2018)
33 Ibid.
34 Horgan (2015)
35 Fortunato et al. (2018)
36 Mallapaty (2018), Chawla (2019)
37 Wang and Evans (2019)
38 Cowen and Southwood (2019)
39 Rockey (2015)
40 Lakatos (1980)
41 See for example Ritchie (2020) for more detail.
42 Baker (2016)
43 Ritchie (2020)
44 See ibid and Ioannidis (2005) for one of the earlier papers that rang the alarm bell.
45 Arbesman (2012), p. 108
46 Smolin (2008), p. 66. Others might say that confirmation of W and Z bosons in 1984 qualifies.
47 Hossenfelder (2018)
48 Cossins (2019)
49 See for example Le Fanu (2010)
50 See Loeb (2021) and Solms (2021) on aliens and consciousness respectively for recent examples. Even these eminent scientists have had to risk their reputations to work on these frontiers.
51 To use the phrase of the former *Nature* editor John Maddox (1999).
52 Mokyr (2018)
53 Brockliss (2019), p. 110
54 https://researchsupport.admin.ox.ac.uk/information/income
55 Brockliss (2019), p. 125
56 See Pinker (2019) or Rosling (2018), or the work of Matt Ridley.
57 Greenspan and Wooldridge (2018), p. 400
58 Pinker (2019), p. 240
59 Ibid., p. 251
60 Ibid.
61 https://www.chronicle.com/article/Which-Colleges-Have-the/245587
62 Lukianoff and Haidt (2018), p. 197
63 Reller (2016)

64 Gastfriend (2015)
65 Jones (2008)
66 National Science Board (2020). The $2.2tn is a 2017 figure and likely conservative for 2021 and beyond.
67 Ibid. and see for example Arond and Bell (2010)
68 OECD (2020)
69 Rosling (2018), p. 62
70 Kelly (2017), p. 284

INTERLUDE

1 Glendon (2002), p. 228. The account here is largely derived from Glendon's.
2 Ibid., p. 166
3 Ibid., p. 171
4 Joseph Henrich (2020) traces such 'natural' rights back to psychological and social changes of the High Middle Ages. See Glendon (2011) for more on Cicero. Other early thinkers about 'natural law' would include the School of Salamanca in Spain, the Dutch diplomat Hugo Grotius or the German political philosopher Samuel von Pufendorf.
5 See for example Robertson (2020); Mokyr (2011).
6 Israel (2006)
7 See for example the work of Samuel Moyn who argues that the modern sense dates from even more recently than the postwar period.

5 THE IDEA PARADOX

1 Clery (2013), p. 21
2 IEA (2019)
3 Quoted on Clery (2013), p. 307
4 Kean (2019)
5 Alexander (2018)
6 The phenomenon of multiple discovery has already been discussed, but I count these as having happened 'once' in historical terms.
7 The metaphor is used extensively in Cowen (2011), for example.
8 This argument is elaborated at length in Gordon (2016).

9 The Greeks discovered the heliocentric model but then forgot about it. Nor was Copernicus alone in his thinking. For example, fifteenth-century astronomers and thinkers like Nicholas of Cusa and Regiomontanus had suggested the Earth was in motion.
10 See for example https://www.archaeology.org/issues/323-1901/features/7196-top-10-discoveries-of-2018
11 There are of course candidates – the eruption of the volcano Thera around the island of Santorini being one.
12 Account taken from Robinson (2006).
13 Jones, Reedy and Weinberg (2014)
14 The following research is taken from Jones (2009).
15 Mulgan (2018), p. 57
16 Wuchty, Jones and Uzzi (2007)
17 Lingfei Wu and Evans (2019)
18 Epstein (2019), p. 180
19 Jones (2008) and Jones, Reedy and Weinberg (2014)
20 Jones (2008)
21 Jones and Weinberg (2011)
22 Jones (2010)
23 Jones and Weinberg (2011)
24 Chai (2017)
25 There is also the more delicate matter, especially relevant in fields like maths and physics, that cognitive performance starts to decline from around thirty. Creativity too may decline. For much more on the complex impact of ageing on new ideas see Ricón (2020).
26 Packalen and Bhattacharya (2019). There is a coda: having a lead young researcher but an older associated researcher was the best of all.
27 Azoulay, Graff-Zivin and Fons-Rosen (2019)
28 Jones and Weinberg (2011)
29 In a call with the author, 14 June 2019.
30 Homer-Dixon (2001)
31 Clearfield and Tilcsik (2018)
32 Arbesman (2017), p. 34
33 Ibid., p. 35
34 See for example Gleick (1988)
35 Quoted in Watson (2017), p. 385
36 Skinner (2016)
37 Mesoudi (2011)

38 Bromham, Dinnage and Hua (2016)
39 See for example Horgan (2015) and Higgs (2019) for more on the challenges.
40 Quoted in Boudry (2019), more generally about the mysterian philosophers like Jerry Fodor and Colin McGinn.
41 Henrich (2016)
42 Arbesman (2017)
43 See Wilson (2017)
44 Rees (2018), p. 193
45 The physicist Seth Lloyd calculates that since the Big Bang the universe has performed 10^{120} operations on 10^{90} bits (Lloyd 2001). In other words, the universe itself has a finite computational capacity.
46 Horgan (2015)
47 Alexander (2018)
48 Scharf (2015)
49 Alexey Guzey (2019) uses a similar explanation for why we are likely disappointed by technological progress.
50 Dorling (2020), p. 208
51 Howes (2020)
52 Scannell et al. (2012)
53 Arbesman (2011)
54 In the words of Geoff Mulgan: 'The study of wisdom in different civilizations and eras has confirmed that there are surprisingly convergent views of what counts as wisdom.' Mulgan (2018), p. 224.
55 Sarewitz (2016)
56 Joyner, Paneth and Ioannidis (2016)
57 See for example Johnson (2011) and research on growth like Weitzman (1996).
58 Fernández-Armesto (2019), p. 401

6 THE STAGNANT SOCIETY

1 Gertner (2012), p. 1
2 Quoted ibid., p. 307
3 Ibid., p. 150
4 Quoted ibid., p. 184
5 Southwood (2020)
6 Odlyzko (1995)
7 Ibid.

8 Arora, Belenzon and Patacconi (2015)
9 As pointed out in Erixon and Weigel (2016)
10 This is for US companies as reported by the National Science Foundation, https://www.nsf.gov/statistics/2019/nsf19326/
11 See for example Goldin and Kutarna (2017), p. 393 or Hockfield (2019). US Federal investment in R&D went from 2 per cent of GDP in the 1960s to under 1 per cent today.
12 Fleming et al. (2019)
13 Arora et al. (2020). It should also be noted that tech is no panacea here and these trends also hold true there.
14 Gruber and Johnson (2019), p. 108
15 Jaffe and Jones (2015), p. 8
16 Ibid. – this is the wider argument of the book.
17 Lucking, Bloom and Van Reenen (2018)
18 A review of the literature on corporate R&D from Ricón (2015) suggests that outsourcing indeed is what happened. However this doesn't address how much of a problem that is. Recent research from Arora et al. (2020) strongly suggests that it has a negative impact.
19 See such names as Vico, Nietzsche, Toynbee, Kondratiev and Huntington, for example.
20 Spengler (1991)
21 *The Economist* (2016)
22 Janik and Toulmin (1973)
23 Mazzucato (2018), p. 177 and p. 174
24 Quoted ibid., p. 175
25 Vollrath (2020), p. 118
26 Haldane (2015)
27 See for example: https://www.ft.com/content/69aa638e-3164-11ea-9703-eea0cae3f0de
28 Faulkender, Hoskins and Petersen (2019)
29 Hopenhayn, Neira and Singhania (2018)
30 Akcigit and Ates (2019)
31 Bannerjee and Duflo (2019), p. 178
32 For a concrete look at how the financialised mindset impacts university research, for example, see Hvide and Jones (2018).
33 Cowen (2018a), p. 79
34 Bessen et al. (2020)
35 Mazzucato (2013)

36 See for example a paper like this for an account of how VC can be detrimental to innovation: Howell et al. (2020)
37 Account from a conversation with Richard A.L. Jones.
38 See for example Gruber and Johnson (2019), p. 100
39 Jaffe and Lerner (2004)
40 Ibid., p. 11
41 See for example Akcigit and Ates (2019). They directly connect the concentration of patents to a wider slowdown in knowledge diffusion and business dynamism.
42 Jaffe and Lerner (2004), p. 2
43 Ibid., p. 3
44 Ibid., pp. 32–4
45 Ibid., p. 58
46 Erixon and Weigel (2016), p. 65
47 Ginsberg (2013)
48 Ibid., p. 35
49 Lukianoff and Haidt (2018), p. 199
50 Muller (2018), p. 139
51 Ginsberg (2013), p. 66
52 Guzey (2019a)
53 Just how bad things like peer review are is debatable. At one end of the scale there are people who think it's bad, but there are no good replacements. At the other are thinkers like Donald W. Braben, who argues that peer review is directly responsible for a collapse in what he calls 'transformative research': 'for the first time since the Renaissance, the limits of thinking began to be systematically curtailed.' Braben (2020), p. 31.
54 Foster, Rzhetsky and Evans (2015)
55 Ibid.
56 Bhattacharya and Packalen (2020)
57 For more on this see for example Chai and Menon (2018). They show the nature of the gamble: the most original work covering rarely addressed points does get the most citations. But the bias against the novelty effect is powerful, and such work is more likely to go unrecognised.
58 Ritchie (2020) has a mass of detail on this process; see also Clancy (2020a).
59 In a conversation with me the director of the Wellcome Trust, Sir

Jeremy Farrar, made all these points. For a concrete study, see Packalen and Bhattacharya (2020) on NIH grant funding.

60 Cook (2011)

61 Boudreau, Guinan, Lakhani and Riedl (2016)

62 Smolin (2008), p. 328

63 For other eminent researchers who believe they would not have got through in the present system see for example Buck (2020).

64 Ginsberg (2013), p. 131

65 See for example Marcus (2018), and also the numbers in Goldin and Kutarna (2017), p. 436.

66 Wellcome Trust (2020)

67 Sverdlov (2018)

68 Morieux (2017)

69 See https://www.bcg.com/en-gb/capabilities/smart-simplicity/complicatedness-survey; Erixon and Weigel (2016), p. 88. The Index looks at 'the amount of procedures, vertical layers, interface structures, coordination bodies, and decision approvals within organizations'.

70 Kirsner (2018)

71 See for example Cummings (2014)

72 Erixon and Weigel (2016), p. 153

73 Storrs Hall (2018)

74 Nichols (2017)

75 See for example O'Mahony (2019), p. 199

76 Drezner (2017)

77 Ibid.

78 Ibid.

79 Lukianoff and Haidt (2018), p. 110

80 Ibid., p. 111

81 Ibid.

82 Bannerjee and Duflo (2019), p. 1

83 Ibid., p. 127

84 O'Connor and Weatherall (2018)

85 Haidt and Lukianoff (2018), p. 77

86 It is curious that many of the pioneers of a previous generation, from Peter Singer to Peter Tatchell, have been 'cancelled' and deplatformed from speaking engagements.

87 Thompson and Smulewicz-Zucker (2018), p. 132

88 Tollefson (2020)

89 See for example Belot (2018)
90 Kaufmann (2010)
91 Drezner (2017)
92 Williams (2018). Alternatively: go on Twitter.
93 Davies (2019)
94 Thanks to Erixon and Weigel (2016) for these two insights.
95 O'Mahony (2019), p. 256
96 See for example Goldin and Kutarna (2017), p. 264
97 O'Mahony (2019), p. 256
98 Erixon and Weigel (2016), p. 142; see also Roy (2012).
99 Grush (2019)
100 Brennan (2019)
101 Storrs Hall (2018)
102 Wallace-Wells (2019), p. 7. Further figures taken from the same source.
103 Lenton et al. (2019)
104 Wallace-Wells (2019), p. 180
105 Ibid., p. 181
106 Rifkin (2014), pp. 82–3
107 Cowen (2018a)
108 Storrs Hall (2018)

7 THE WORLD'S NEW TOOLKIT

1 Goldin and Kutarna (2017), p. 186
2 Senior et al. (2020)
3 Ibid.
4 The founder of CASP prefers to call it an experiment rather than a competition, even if the competitive edge is a large part of what makes it work.
5 AlQuraishi (2018)
6 Reynolds (2020)
7 McAfee and Brynjolfsson (2017), p. 2
8 Callaway (2020)
9 AlQuraishi (2020)
10 Hassabis (2019)
11 Wootton (2015)
12 Ibid.
13 Eisenstein (1979)

14 Wootton (2015), p. 215
15 Although the discovery of sunspots seems to be another example of multiple discovery, also found at the same time by astronomers in Oxford, Ingolstadt and Wittenberg.
16 Galileo quoted in Koestler (1964), p. 336
17 Wootton (2015), p. 236
18 This point is widely made – see for example Mulgan (2018), Agar (2012) or Ridley (2016) for wider discussions.
19 Tools are not then sufficient conditions but they are (often) necessary conditions.
20 https://hbr.org/podcast/2020/10/deepminds-journey-from-games-to-fundamental-science
21 https://home.cern/science/computing/processing-what-record
22 Goldin and Kutarna (2017), p. 241
23 Ringel et al. (2020)
24 Stokes (2020)
25 McMahon (2020)
26 Even that is quickly dated: since releasing GPT-3 Google has a model with trillions of parameters, and the numbers will keep growing.
27 Tshitoyan et al. (2019)
28 Rotman (2019)
29 Malone (2018), p. 240
30 Du Sautoy (2019), Miller (2019)
31 Hafner et al. (2019)
32 Quoted in Parker (2020)
33 Assael, Sommerschield and Prag (2019)
34 Each is a genuine application of AI and not just a series of cool things.
35 See Kelly (2017), pp. 45–6 for a taxonomy of possible minds.
36 Bostrom (2017)
37 Lovelock (2019)
38 Experiments on B-mesons and Muon g-2 respectively. It is still early days as of writing.
39 Watney (2020)
40 Vince (2019), p. 231
41 Goldin and Kutarna (2017), p. 236
42 Martinis and Boixo (2019)
43 My thanks to Vera Schäfer of the University of Oxford Physics Department for talking me through how this works.

44 Drexler (2013)
45 See Hockfield (2019) and Contera (2019)
46 Carey (2019), p. 13
47 Although it is worth saying that this was based on much earlier, highly exploratory work elsewhere – another good reason why incentives that take us away from exploratory science are a worrying sign.
48 Quoted in Carey (2019), p. 142
49 Conde, Pande and Yoo (2019)
50 To get a sense of how fast things are developing see Meng and Ellis (2020). Over the past decade researchers have achieved full synthesis of *E.coli* bacterium.
51 See for example Hockfield (2019) or Morton (2019).
52 McKinsey Global Institute (2020)
53 Hanson (2016)
54 Ibid.
55 See for example ibid or Bostrom (2017), Tegmark (2017) and Lovelock (2019).
56 https://app.ft.com/cms/s/9e5abb2a-7deb-40ed-a0fc-4b24b4458445.html?sectionid=tech
57 Carey (2020)
58 Thompson et al. (2020)
59 Klinger, Mateos-Garcia and Stathoulopoulos (2020)
60 Malone (2018), p. 65
61 Martin (2015)
62 Gordon (2016)
63 See the extraordinary website https://wtfhappenedin1971.com/ for more on this change.
64 Perez (2002)
65 In fairness, even the proponents of the 4IR acknowledge there is a debate. Klaus Schwab of the World Economic Forum believes that an increased velocity of technical change does justify the 4IR (Schwab 2017) although I disagree.
66 Again, Kelly (2017) makes this point: imagine looking at the 3IR from the vantage of 2095: any time between 1980 and 2050 could arguably be seen as in motion and its full impact not possible to judge from that position.
67 A schema outlined in Kelly (2016).

8 THE GREAT CONVERGENCE

1 Cyranoski (2019)
2 Although, controversially, scientists at places like Stanford and Rice universities had been in the know.
3 Ross (2017), p. 67
4 Carey (2019), p. 86
5 The phrase 'great convergence' is Richard Baldwin's.
6 Khanna (2019), p. 72
7 DeLong (2018)
8 Morris (2010)
9 Ibid., p. 591
10 Goldin and Kutarna (2017), p. 62
11 Ibid., p. 64
12 National Science Board (2020)
13 Gruber and Johnson (2019), p. 214
14 Okoshi (2019)
15 Ibid., p. 204; National Science Board (2020)
16 National Science Board (2020)
17 Khanna (2019), pp. 199–200
18 Xinhua (2019)
19 Khanna (2019), p. 148
20 Goldin and Kutarna (2017), p. 64
21 Ibid., p. 164
22 Khanna (2019), p. 211
23 Goldin and Kutarna (2017), p. 125
24 Ibid., p. 164
25 Ibid., p. 154
26 Vollset et al. (2020)
27 Urquiola (2020)
28 Gruber and Johnson (2019), p. 34
29 Rosling (2018)
30 Dorling (2020), p. 276
31 Rosling (2018)
32 Goldin and Kutarna (2017), p. 391
33 https://www.timeshighereducation.com/data-bites/data-bite-share-female-professors-now-virtually-quarter
34 https://www.catalyst.org/research/women-in-academia/
35 Teare (2020)

36 Bannerjee and Duflo (2019), p. 167
37 Moretti (2019)
38 Smil (2020), pp. 44–9
39 Freedman (2018), p. 255
40 For an example of this see Saudi Arabia's The Line. Certainly a big idea for a city, whether a good one time will tell.
41 And historical analysis also supports this. See for example Henrich (2020), pp. 448–52
42 Packalen and Bhattacharya (2015)
43 Berkes and Gaetani (2019)
44 Ibid.
45 Schwab (2017)
46 Smith (2021)
47 Coggan (2020), p. 58
48 Hockfield (2019), p. 163
49 Hunt and Gauthier-Loiselle (2010)
50 This is generally not appreciated often enough. In the words of Cowen and Southwood (2019): 'Progress on the aggregates but not the per capita rates of change seems to reflect quite significant and indeed somewhat neglected features of the contemporary world.'
51 Kremer (1993)
52 Henrich (2009), Muthukrishna and Henrich (2016)
53 Daxx (2020)
54 See for example Singh and Fleming (2010) or Hargadon (2003).
55 For more discussion of this model of innovation see for example Howes (2021).
56 Robertson (2020), p. 374
57 Figure taken from Norberg (2020), p. 157
58 Thirty million is a very rough estimate here, but given there are many more than that around the world with a postgraduate degree, it doesn't seem an unreasonable number for highly educated, intelligent and engaged global citizens. It could even be a colossal underestimate.
59 Agarwal and Gaulé (2018)
60 And even if only a tiny number ever originate or deliver big ideas, there should be more in total at the frontier.
61 Numbers here are taken from Reinhardt (2020)
62 See Henrich (2020), p. 463 for a review of the extensive literature on this point.

63 See for example Bloom, Van Reenen and Williams (2019) but also Mokyr (2017), p. 123, even if it is increasingly obvious education and human capital formation more widely isn't a panacea.
64 Coggan (2020), p. 290
65 World Bank education report http://www.worldbank.org/en/publication/wdr2018
66 Watson (2001)
67 Brockliss (2019), p. 108
68 Ridley (2011), p. 97
69 Taken from https://ourworldindata.org/ under Creative Commons licence.
70 Sargent (2020)
71 Hawksworth, Clarry and Audino (2017)
72 See Clancy (2021a) and (2021b) for a review of this literature.
73 Evans and Lambert (2020)
74 My thanks to Ian Goldin for suggesting this framing.
75 See for example Syed (2019)
76 Ridley (2011) and (2016)
77 Malone (2018)
78 Rogers Hollingsworth (2007)
79 Extensively detailed in Nisbett (2003) and Henrich (2020)
80 Vince (2019), p. 179
81 Mokyr (1994)
82 Bannerjee and Duflo (2019)
83 Vollset et al. (2020)
84 Ridley (2020)
85 Gruber and Johnson (2019), p. 19
86 Wuchty, Jones and Uzzi (2007)
87 Douglas (1986), p. 8
88 There is a long literature about institutions which looks at them specifically in their wider societal sense. Here I am looking at that meaning, but also at the tighter and more localised version which overlaps with the term organisation.
89 See for example Furman and Stern (2011) for a specific study. More generally the oeuvre of economists like Douglass North or Joel Mokyr is built around this insight to the extent that it has in various forms become a kind of foundational tenet of the social sciences.
90 See for example Malone (2018) or Mulgan (2018).

91 Mulgan (2018), p. 56
92 Dowey (2017)
93 See for example Mokyr (2017), p. 142: 'What changed in this age was the culture – the beliefs and attitudes of the educated elite toward useful knowledge, how to acquire it, how to distribute it, what it could do. Such changing beliefs led to new institutions reflecting them, and those institutions fed back into the beliefs.'
94 McCormick (2020)
95 Some are beginning the process of designing new types of organisation. See for example Rodriques and Marblestone (2020) on the notion of Focused Research Organisations to overcome the limitations of existing forms, or Benjamin Reinhardt's proposal for a new private sector ARPA (Reinhardt 2021).
96 Verhoeven (2020)
97 Singh and Fleming (2010)
98 If the research here seems contradictory then that is because it is. As Wuchty, Jones and Uzzi (2007) show, however, big teams are both dominant and responsible for most of the high-impact work around. The question is whether this can overcome the impact of loss of disruption as per Lingfei Wu and Evans (2019). Scaling up implies that the world can support the ongoing escalation in difficulty of ideas/team size.
99 Mokyr (2018)
100 Henrich (2016) and Malone (2018)
101 See Henrich (2016) for a substantial demonstration that, as simple as they may seem, ancient tools would and could not have been made by one mind.
102 Vince (2019), p. xvi
103 Kelly (2017)
104 Vince (2019)
105 Kelly (2017), p. 291
106 See Henrich (2020) for a detailed account of how this collective brain scaled up from the early Middle Ages onwards.

9 THE UNCERTAIN HORIZON

1 Used in the sense of Schwartz (1991).
2 Freeman (2003)

3 Quoted in ibid., p. 275
4 Ovenden (2020), pp. 34–7
5 Whitehead (1925), p. 7
6 It is important to point out that this was true only of science or propositional knowledge. In terms of technology or procedural knowledge the Middle Ages were a remarkably fecund period and exceeded those of classical antiquity (Mokyr 1990).
7 Morris (2010)
8 From a survey by Michael Shermer cited in Cowen and Southwood (2019), p. 31
9 Morris (2010), p. 613. More recent estimates say there 'should' (of course heavily caveated) be far fewer.
10 There have been recent suggestions from for example Sandberg, Drexler and Ord (2018) that the paradox isn't a paradox and there is unlikely to be a great filter. However if there is no Great Filter this means we are likely alone, and the weight of responsibility just as great.
11 See for example comments from Slavoj Žižek, William Gibson or Andre Spicer.
12 Here taken from Howes (2019). Others like Henrich (2020) try the thought experiment from earlier (the year 1000 CE for example), where it is even more unflattering to England.
13 Watney (2020), Maçães (2021)
14 Although worth saying that currently it is AI approaches that are making the running in AI – including a repurposed version of DeepMind's AlphaZero game-playing system.
15 Gruber and Johnson (2019), p. 4
16 Ibid., p. 6
17 Ibid., p. 48
18 Ibid., p. 49
19 Mazzucato (2021) p. 83
20 Ibid., p. 55
21 Gross and Sampat (2020)
22 Azoulay and Li (2020)
23 Mazzucato (2018a) and Mazzucato (2021)
24 For a fascinating thought experiment here, see Hooper (2021). It may be that vast sums of money would be the best way of realising many big ideas …
25 Gruber and Johnson (2019), p.122

26 This is a flourishing area; see for example Ioannidis (2018). In particular I'm interested in thinkers outside academia like Alexey Guzey, Adam Marblestone, Michael Nielsen and José Luis Ricón who have a huge range of creative suggestions on how to improve research. See also Armstrong (2019).
27 Azoulay (2012)
28 Bhattacharya and Packalen (2020)
29 Guzey (2019)
30 Azoulay, Graff Zivin and Manso (2011). For a detailed review of the funding of people and not projects see Ricón (2020a).
31 For example Matt Clifford also powerfully made this point, and the organisation he co-founded, Entrepreneur First, is an excellent example in the startup world.
32 Szell, Ma and Sinatra (2018)
33 See for example Mulgan (2018), p. 55
34 Caplan (2018)
35 Ridley (2016), p. 180
36 Cummings (2013)
37 Biasi and Ma (2020). The authors also document the deleterious impact of this gap on those from poorer backgrounds.
38 Jimmy Wales did not attend a Montessori school, but his education was heavily influenced by their principles.
39 Weiman (2017)
40 Ricón (2019)
41 Ibid.
42 Merton (1996)
43 Beck (1992)
44 See for example Armstrong (2020)
45 See for example Bloom, Van Reenen and Williams (2019) or Hvide and Jones (2018). There is a huge literature on what might help boost innovation, and no end of scope to make it better.
46 Thank you to Matt Clifford for making this point to me.
47 And thank you to Carlota Perez for this point.
48 Rogers Hollingsworth (2007)
49 Syed (2019), p. 141
50 Shi and Evans (2019)
51 Uzzi et al. (2013)
52 Merton and Barber (2003)
53 Braben (2020) would dispute this: if you choose your efforts

carefully enough, there is little chance of failure, he would argue; it's just that we don't try properly.

54 Garde and Saltzman (2020)

EPILOGUE

1 Kardashev (1964)

2 Kaku (2012)

Acknowledgements

This book started over six years ago, just as I finished the first draft of my previous project *Curation.* I didn't expect this one to take so long, nor to change as much as it did, or that it would cover such a wide territory. Probably for the best that I didn't!

Many, many people have been involved in countless ways. As a publisher myself, I am keenly aware of what a team effort the production of any book is, this one more than most. My brilliant literary agent Sophie Lambert was instrumental. Many early proposals were politely batted back until it had taken a form which she, quite rightly, thought would work. When I heard my editor Tim Whiting was establishing a new ideas-driven imprint, the Bridge Street Press, I immediately hoped he would be interested. Luckily he was, and he has done an amazing job of shaping the book, from title and subtitle to the overall approach. It simply wouldn't be close to the same project without their guidance. Thanks also to Holly Harley for her countless improvements to the manuscript, Zoe Gullen for guiding things through editorial, to the copy-editor Steve Gove, whose magnificent work has made this a much better book, and to everyone at Little, Brown for making it come together. I'd also like to thank Anne-Marie Bono and the team at MIT Press for their vote of confidence and their fantastic work in Canada and the US, and Rocío Martínez Velázquez and everyone at the Fondo de Cultura Económica in Mexico for their work on the book in Spanish, and their support for me over many years.

One of the great positives about a project like this is that it gave me licence to talk to a wide variety of the world's most interesting and original thinkers. They have been incredibly generous. I'd like to thank the following for chatting generally at an early stage, having a coffee on a specific point, answering my pestering emails once the pandemic hit, or even just forwarding a PDF or a suggestion of further reading.

My thanks to: Euan Adie, Azeem Azhar, Courtney Biles, Francis Casson, Krishan Chadha, Sen Chai, Ben Chamberlain, Tom Chatfield, Harry Cliff, Matt Clifford, Daniel Crewe, Lee Cronin, Payel Das, Danny Dorling, Eric Drexler, Fredrik Erixon, Jeremy Farrar, Iason Gabriel, Ian Goldin, Robert J. Gordon, the late and much missed David Graeber, Alexey Guzey, Anton Howes, William Isaac, Matthew Jockers, Benjamin F. Jones, Richard A.L. Jones, Victoria Krakovna, Roman Krznaric, François Lafond, James Le Fanu, Joel Mokyr, Geoff Mulgan, Mikko Packalen, Carlota Perez, Mark Piesing, Benjamin Reinhardt, Matt Ridley, Jack Scannell, Vera Schäfer, Ben Southwood, Peter Watson and Michael Webb. I'm sure there are many names I've forgotten and apologies in advance to them. A particular thanks to Anthony Blake, James Bullock, Angus Phillips, Daniel Slater, Mustafa Suleyman and George Walkley for their comments on an early draft of the book and for many conversations around it. It goes without saying that all errors are absolutely my own and I welcome readers who find them!

Compared to my previous books I'm also keenly aware this was a discussion being held on blogs, newsletters and podcasts. Among those I've particularly found valuable in writing this book are *Age of Invention*, *Conversations with Tyler*, *Exponential View*, *Ideas Machines*, *Marginal Revolution*, *Narratives*, *New Things Under the Sun*, *Nintil*, *Noahpinion*, *Radical Science*, *Talking Politics*, *The 80,000 Hours Podcast* and *Thoughts in Between*. And, God help me, Twitter. I'd also like to thank all my colleagues at Canelo who are on the journey of a burgeoning start-up, and in particular my co-founders Iain Millar and Nick Barreto for their constant support through all the ups and downs, and to Mustafa and the team for having me at DeepMind.

Lastly the biggest thanks go as always to my wife Danielle, not least given that during the long course of this book not one but two children, to whom it is dedicated, arrived (which is perhaps why it took so long . . .). Thank you for everything.

Bibliography

Agar, Jon (2012), *Science in the Twentieth Century and Beyond*, Cambridge: Polity Press

Agarwal, Ruchir, and Gaulé, Patrick (2018), *Invisible Geniuses: Could the Knowledge Frontier Advance Faster?*, IZA Institute of Labour Economics Discussion Paper

Akcigit, Ufuk, and Ates, Sina T. (2019), 'What Happened to U.S. Business Dynamism?', NBER Working Paper 25756

Alexander, Scott (2018), 'Is Science Slowing Down?', *SlateStarCodex*, accessed 22 October 2020, available at https://slatestarcodex.com/2018/11/26/is-science-slowing-down-2/

AlQuraishi, Mohammed (2018), 'AlphaFold @ CASP13: "What just happened?"', accessed 5 February 2020, available at https://moalquraishi.wordpress.com/2018/12/09/alphafold-casp13-what-just-happened/

AlQuraishi, Mohammed (2020), 'AlphaFold2 @ CASP14: "It feels like one's child has left home"', accessed 23 December 2020, available at https://moalquraishi.wordpress.com/2020/12/08/alphafold2-casp14-it-feels-like-ones-child-has-left-home/

Andersen, Kurt (2011), 'You Say You Want a Devolution?', *Vanity Fair*, accessed 6 October 2020, available at https://www.vanityfair.com/style/2012/01/prisoners-of-style-201201

Arbesman, Samuel (2011), 'Quantifying the Ease of Scientific Discovery', *Scientometrics*, Vol. 86 No. 2, pp. 245–50

Arbesman, Samuel (2012), *The Half-Life of Facts: Why Everything We Know Has An Expiration Date*, New York: Current

Arbesman, Samuel (2017), *Overcomplicated: Technology at the Limits of Comprehension*, New York: Portfolio Penguin

Armstrong, Bryan (2019), 'Ideas on how to improve scientific research', accessed 14 June 2020, available at https://medium.com/@barmstrong/ideas-on-how-to-improve-scientific-research-9e2e56474132

Armstrong, Harry (2020), 'Innovation-enabling approaches to regulation', *Nesta*, accessed 9 July 2020, available at https://www.nesta.org.uk/blog/innovation-enabling-approaches-regulation/

Arond, E. and Bell, M. (2010), *Trends in the Global Distribution of R&D Since the 1970s: Data, their Interpretation and Limitations*, STEPS Working Paper 39, Brighton: STEPS Centre

Arora, Ashish, Belenzon, Sharon and Patacconi, Andrea (2015), 'Killing the Golden Goose? The Decline of Science in Corporate R&D', NBER Working Paper No. w20902

Arora, Ashish, Belenzon, Sharon, Patacconi, Andrea, and Suh, Jungkyu (2020), 'The Changing Structure of American Innovation: Some Cautionary Remarks for Economic Growth', *Innovation Policy and the Economy*, Vol. 20

Asimov, Isaac (2016), *Foundation* (first published 1951), London: HarperVoyager

Assael, Yannis, Sommerschield, Thea, and Prag, Jonathan (2019), 'Restoring ancient text using deep learning: a case study on Greek epigraphy', arXiv, 1910.06262

Azoulay, Pierre (2012), 'Turn the scientific method on ourselves', *Nature*, 484, pp. 31–2

Azoulay, Pierre, and Li, Danielle (2020), 'Scientific Grant Funding', NBER Working Paper 26889

Azoulay, Pierre, Graff Zivin, Joshua S., and Fons-Rosen, Christian (2019), 'Does Science Advance One Funeral at a Time?', *American Economic Review*, Vol. 109 No. 8, pp. 2889–2920

Azoulay, Pierre, Graff Zivin, Joshua S., and Manso, Gustavo (2011), 'Incentives and creativity: evidence from the academic life sciences', *The RAND Journal of Economics*, Vol. 42, No. 3 pp. 527–54

Bahcall, Safi (2019), *Loonshots: How to Nurture the Crazy Ideas that Win Wars, Cure Diseases, and Transform Industries*, New York: St Martin's Press

Baker, Monya (2016), '1,500 scientists lift the lid on reproducibility', *Nature* 533, 452–4

Banerjee, Abhijit V., and Esther Duflo (2019), *Good Economics for Hard Times: Better Answers to Our Biggest Problems*, London: Allen Lane

Beck, Ulrich (1992), *Risk Society: Towards a New Modernity*, London: SAGE

Belot, Henry (2018), 'Nobel Prize winner Peter Doherty criticises national interest test on research funding', *ABC*, accessed 31 October 2018, available at https://amp.abc.net.au/article/10450504

Berkes, Enrico and Gaetani, Ruben (2019), 'The Geography of Unconventional Innovation', Rotman School of Management Working Paper No. 3423143

Bessen, James E., Denk, Erich, Kim, Joowon, and Righi, Cesare (2020), 'Declining Industrial Disruption', Boston Univ. School of Law, Law and Economics Research Paper 20–28

Bhattacharya, Jay, and Packalen, Mikko (2020), 'Stagnation and Scientific Incentives', NBER Working Paper 26752, available at https://www.nber.org/papers/w26752

Biasi, Barbara, and Ma, Song (2020), 'The Education-Innovation Gap', accessed 15 January 2021, available at https://songma.github.io/files/bm_edu_inno_gap.pdf

Bikard, Michaël (2020), 'Idea Twins: Simultaneous Discoveries as a Research Tool', *Strategic Management Journal*, Vol. 41 No. 8, pp. 1528–43

Bloom, Nicholas, Jones, Charles I., Van Reenen, John, and Webb, Michael (2020), 'Are Ideas Getting Harder to Find?', *American Economic Review*, Vol. 110 No. 4, pp. 1104–44

Bloom, Nicholas, Van Reenen, John, and Williams, Heidi (2019), 'A Toolkit of Policies to Promote Innovation', *Journal of Economic Perspectives*, Vol. 33 No. 3, pp. 163–84

Bostrom, Nick (2017), *Superintelligence: Paths, Strategies, Dangers*, Oxford: Oxford University Press

Boudreau, Kevin, Guinan, Eva, Lakhani, Karim R., and Riedl,

Christoph (2016), 'Looking Across and Looking Beyond the Knowledge Frontier: Intellectual Distance and Resource Allocation in Science', *Management Science*, Vol. 62 No. 10, pp. 2765–3084

Boudry, Maarten (2019), 'Human intelligence: have we reached the limit of knowledge?', The Conversation, accessed 29 October 2019, available at https://theconversation.com/human-intelligence-have-we-reached-the-limit-of-knowledge-124819

Braben, Donald W. (2020), *Scientific Freedom: The Elixir of Civilization*, San Francisco: Stripe Press

Brennan, Reilly (2019), 'The State of Autonomous Transportation', *Exponential View*, accessed 11 January 2021, available at https://www.exponentialview.co/p/the-state-of-autonomous-transportation

Brockliss, Laurence (2019), *The University of Oxford: A Brief History*, Oxford: Bodleian Library

Brockman, John (ed.) (2020), *Possible Minds: 25 Ways of Looking At AI*, New York: Penguin

Bromham, Lindell, Dinnage, Russell, and Hua, Xia (2016), 'Interdisciplinary research has consistently lower funding success', *Nature* 534, pp. 684–7

Buck, Stuart, 'Escaping Science's Paradox', *Works in Progress*, accessed 7 April 2021, available at https://worksinprogress.co/issue/escaping-sciences-paradox/

Burke, Peter (2020), *The Polymath: A Cultural History from Leonardo da Vinci to Susan Sontag*, New Haven, CT: Yale University Press

Bush, Vannevar (1945), *Science: The Endless Frontier*, Washington DC: United States Government Printing Office

Callard, Agnes (2020), 'Publish and Perish', The Point, accessed 6 October 2020, available at https://thepointmag.com/examined-life/publish-and-perish-agnes-callard/

Callaway, Ewen (2020), '"It will change everything": DeepMind's AI makes gigantic leap in solving protein structures', *Nature*, accessed 23 December 2020, available at https://www.nature.com/articles/d41586-020-03348-4

Campbell, Chris, and Mathurin, Patrick (2020), 'Hollywood "sequelitis"', *Financial Times*, accessed 9 March 2020, available at https://www.ft.com/content/6d5871d8-3ea7-11ea-b232-000f4477fbca

Cancer Research UK (2019), 'Worldwide cancer statistics', accessed 20 April 2019, available at https://www.cancerresearchuk.org/health-professional/cancer-statistics/worldwide-cancer#heading-Zero

Caplan, Bryan (2018), *The Case against Education: Why the Education System Is a Waste of Time and Money*, Princeton: Princeton University Press

Carey, Nessa (2019), *Hacking the Code of Life: How Gene Editing Will Rewrite Our Futures*, London: Icon Books

Carey, Ryan (2020), 'Interpreting AI compute trends', *AI Impacts*, accessed 11 January 2020, available at https://aiimpacts.org/interpreting-ai-compute-trends/

Chai, Sen (2017), 'Near Misses in the Breakthrough Discovery Process', *Organization Science*, Vol. 28 No. 3, pp. 411–28

Chai, Sen, and Menon, Anoop (2018), 'Breakthrough recognition: Bias against novelty and competition for attention', *Research Policy*, Vol. 48 No. 3, pp. 733–47

Chawla, Dalmeet Singh (2019), 'Hyperauthorship: global projects spark surge in thousand-author papers', *Nature*, accessed 29 March 2021, available at https://www.nature.com/articles/d41586-019-03862-0

Christensen, Clayton M. (2013), *The Innovator's Dilemma: When New Technologies Cause Great Firms To Fail* (reprint edition), Cambridge, MA: Harvard Business Review Press

Clancy, Matt (2020a), 'How bad is publish-or-perish for the quality of science?', *New Things Under The Sun*, accessed 12 January 2021, available at https://mattsclancy.substack.com/p/how-bad-is-publish-or-perish-for

Clancy, Matt (2020b), 'What ails the social sciences', *Works in Progress*, accessed 12 January 2021, available at https://worksinprogress.co/issue/what-ails-the-social-sciences/

Clancy, Matt (2021a), 'More Science Leads to More Innovation', *New Things Under The Sun*, accessed 12 April 2021, available at https://mattsclancy.substack.com/p/more-science-leads-to-more-innovation

Clancy, Matt (2021b), 'Ripples in the River of Knowledge', *New Things Under The Sun*, accessed 12 April, available at https://mattsclancy.substack.com/p/ripples-in-the-river-of-knowledge

Clearfield, Chris, and Tilcsik, András (2018), *Meltdown: Why Our Systems Fail and What We Can Do About It*, London: Atlantic Books

Clery, Daniel (2013), *A Piece of the Sun: The Quest for Fusion Energy*, London: Duckworth Overlook

Coggan, Philip (2020), *More: The 10,000 Year Rise of the World Economy*, London: Economist Books

Collier, Paul (2018), *The Future of Capitalism: Facing the New Anxieties*, London: Allen Lane

Collison, Patrick, and Nielsen, Michael (2018), 'Science Is Getting Less Bang for Its Buck', *The Atlantic*, accessed 7 December 2018, available at https://www.theatlantic.com/science/archive/2018/11/diminishing-returns-science/575665/

Colvile, Robert (2017), *The Great Acceleration: How the World Is Getting Faster, Faster*, London: Bloomsbury

Conde, Jorge, Pande, Vijay, and Yoo, Julie (2019), 'Biology is Eating the World: A Manifesto', a16z, accessed 15 January 2020, available at https://a16z.com/2019/10/28/biology-eating-world-a16z-manifesto/

Conference of the Institute for the Study of Free Enterprise Systems, pp. S71–S102

Contera, Sonia (2019), *Nano Comes to Life: How Nanotechnology is Transforming Medicine and the Future of Biology*, Princeton, NJ: Princeton University Press

Cook, John D. (2011), 'How much time do scientists spend chasing grants?', accessed 25 August 2019, available at https://www.johndcook.com/blog/2011/04/25/chasing-grants/

Cossins, Daniel (2019), '"We'll die before we find the answer": Crisis at the heart of physics', *New Scientist*, 18 September 2019, available at https://www.newscientist.com/article/mg24132130-600-well-die-before-we-find-the-answer-crisis-at-the-heart-of-physics/

Cowen, Tyler (2011), *The Great Stagnation: How America Ate All the Low-Hanging Fruit of Modern History, Got Sick, and Will (Eventually) Feel Better*, New York: Dutton

Cowen, Tyler (2018a), *The Complacent Class: The Self-Defeating Quest for the American Dream*, New York: Picador

Cowen, Tyler (2018b), *Stubborn Attachments: A Vision for a Society of Free, Prosperous, and Responsible Individuals*, San Francisco: Stripe Press

Cowen, Tyler, and Southwood, Ben (2019), 'Is the rate of scientific progress slowing down?', accessed 21 October 2020, available at https://docs.google.com/document/d/1cEBsj18Y4NnVx5Qdu43cKEHMaVBODTTyfHBa8GIRSec/

Cummings, Dominic (2013), 'Some thoughts on education and political priorities', accessed 20 June 2020, available at https://dominiccummings.files.wordpress.com/2013/11/20130825-some-thoughts-on-education-and-political-priorities-version-2-final.pdf

Cummings, Dominic (2014), 'The Hollow Men II: Some reflections on Westminster and Whitehall dysfunction', accessed 8 January 2019, available at https://dominiccummings.com/2014/10/30/the-hollow-men-ii-some-reflections-on-westminster-and-whitehall-dysfunction/

Cyranoski, David (2019), 'The CRISPR-baby scandal: what's next for human gene-editing', *Nature*, accessed 15 March 2020, available at https://www.nature.com/articles/d41586-019-00673-1

Dasandi, Niheer (2018), *Is Democracy Failing? A primer for the 21st century*, London: Thames & Hudson

Davies, William (2019), *Nervous States: How Feeling Took Over the World*, London: Vintage

Daxx (2020), 'How Many Software Developers Are in the US and the World?', accessed 30 May 2020, available at https://www.daxx.com/blog/development-trends/number-software-developers-world

Decker, Ryan, Haltiwanger, John, Jarmin, Ron, and Miranda, Javier (2014), 'The Role of Entrepreneurship in US Job Creation and Economic Dynamism', *Journal of Economic Perspectives*, Vol. 28 No. 3, pp. 3–24

DeLong, Bradford, 'Why Was the 20th Century Not a Chinese

Century?', accessed 9 October 2018, available at https://www.bradford-delong.com/2018/07/why-was-the-20th-century-not-a-chinese-century-an-outtake-from-slouching-towards-utopia-an-economic-history-of-the-long.html

Di Pasquale, Giovanni (ed.) (2018), *Archimedes In Syracuse*, Milan: Giunti Editore

Diamond, Jared (2005), *Collapse: How Societies Choose to Fail or Survive*, London: Penguin

Dorling, Danny (2020), *Slowdown: The End of the Great Acceleration – And Why It's Good for the Planet, the Economy and Our Lives*, New Haven, CT: Yale University Press

Douglas, Mary (1986), *How Institutions Think*, Syracuse, NY: Syracuse University Press

Douthat, Ross (2020), *The Decadent Society: How We Became the Victims of Our Own Success*, New York: Avid Reader Press

Dowey, James (2017), *Mind Over Matter: Access to Knowledge and the British Industrial Revolution*, London School of Economics dissertation

Drexler, K. Eric (2013), *Radical Abundance: How A Revolution in Nanotechnology Will Change Civilization*, New York: Public Affairs

Drezner, Daniel W. (2017), *The Ideas Industry: How Pessimists, Partisans and Plutocrats are Transforming the Marketplace of Ideas*, New York: Oxford University Press

Dyson, Freeman (2006), *The Scientist as Rebel*, New York: New Review of Books Press

Eagleton, Terry (2003), *After Theory*, London: Allen Lane

Edgerton, David (2019), *The Shock of the Old: Technology and Global History Since 1900* (revised edition), London: Profile Books

Eisenstein, Elizabeth L. (1979), *The Printing Press as an Agent of Change: Communications and Cultural Transformations in Early-Modern Europe*, Cambridge: Cambridge University Press

Epstein, David (2019), *Range: How Generalists Triumph in a Specialized World*, London: Macmillan

Erixon, Frederik, and Björn Weigel (2016), *The Innovation Illusion:*

How So Little Is Created By So Many Working So Hard, New Haven, CT: Yale University Press

Evans, Benedict, and Lambert, Oliver (2020), 'Europe, Unicorns and Global Tech Diffusion – The End of the American Internet', *Mosaic Ventures*, accessed 9 January 2021, available at https://www.mosaicventures.com/patterns/europe-unicorns-and-global-tech-diffusion-the-end-of-the-american-internet

Faulkender, Michael W., Hankins, Kristine W., and Petersen, Mitchell A. (2019), 'Understanding the Rise in Corporate Cash: Precautionary Savings or Foreign Taxes', *The Review of Financial Studies*, Vol. 32 No. 9, pp. 3299–3334

Fernández-Armesto, Felipe (2019), *Out of our Minds: What We Think and How We Came to Think It*, London: Oneworld

Fishman, Charles (2019), *One Giant Leap: The Impossible Mission that Flew Us to the Moon*, New York: Simon & Schuster

Fleming, L., Greene, H., Li, G., Marx, M., and Yao, D. (2019), 'Government-funded research increasingly fuels innovation', *Science*, 364, pp. 1139–41

Fortunato, Sandro et al. (2018), 'Science of science', *Science*, Vol. 359 No. 6379

Foster, Jacob G., Rzhetsky, Andrey, and Evans, James A. (2015), 'Tradition and Innovation in Scientists' Research Strategies', *American Sociological Review*, Vol. 80 No. 5

Franklin, Daniel (ed.) (2017), *Megatech: Technology in 2050*, London: Economist Books

Franklin, Daniel, with John Andrews (eds) (2012), *Megachange: The World in 2050*, London: Economist Books

Frase, Peter (2016), *Four Futures: Life After Capitalism*, London: Verso

Freedman, Lawrence (2018), *The Future of War: A History*, London: Penguin

Freeman, Charles (2003), *The Closing of the Western Mind: The Rise of Faith and the Fall of Reason*, London: Pimlico

Fukuyama, Francis (1992), *The End of History and The Last Man*, London: Hamish Hamilton

Funk, Russell, and Owens-Smith, Jason (2017), 'A Dynamic

Network Measure of Technological Change', *Management Science*, Vol. 63 No. 3, pp. 791–817

Furman, Jeffrey L., and Stern, Scott (2011), 'Climbing atop the Shoulders of Giants: The Impact of Institutions on Cumulative Research', *American Economic Review*, Vol. 101 No. 5, pp. 1933–63

Garde, Damian, and Saltzman, Jonathan (2020), 'The story of mRNA: How a once-dismissed idea became a leading technology in the Covid vaccine race', Stat, accessed 23 December 2020, available at https://www.statnews.com/2020/11/10/the-story-of-mrna-how-a-once-dismissed-idea-became-a-leading-technology-in-the-covid-vaccine-race/

Gastfriend, Eric (2015), '90% of all the scientists that ever lived are alive today', Future of Life Institute, accessed 27 July 2019, available at https://futureoflife.org/2015/11/05/90-of-all-the-scientists-that-ever-lived-are-alive-today/

Gay, Peter (2009), *Modernism: The Lure of Heresy, From Baudelaire to Beckett and Beyond*, London: Vintage

Gertner, Jon (2012), *The Idea Factory: Bell Labs and the Great Age of American Innovation*, New York: Penguin

Ginsberg, Benjamin (2013), *The Fall of the Faculty: The Rise of the All-Administrative University and Why It Matters*, New York: Oxford University Press

Gladwell, Malcolm (2008), 'In the Air: Who says big ideas are rare?', *New Yorker*, accessed 19 January 2016, available at https://www.newyorker.com/magazine/2008/05/12/in-the-air

Gleick, James (1988), *Chaos: The Amazing Science of the Unpredictable*, London: Vintage

Glendon, Mary Ann (2002), *A World Made New: Eleanor Roosevelt and the Universal Declaration of Human Rights*, New York: Random House

Glendon, Mary Ann (2011), *The Forum and The Tower: How Scholars and Politicians Have Imagined the World, from Plato to Eleanor Roosevelt*, Oxford: Oxford University Press

Goldin, Ian, and Chris Kutarna (2017), *Age of Discovery: Navigating the Storms of Our Second Renaissance*, London: Bloomsbury Business

Gordon, Robert J. (2016), *The Rise and Fall of American Growth: The U.S. Standard of Living Since the Civil War*, Princeton, NJ: Princeton University Press

Gould, Stephen Jay (2007), *Punctuated Equilibrium*, Cambridge, MA: The Belknap Press of Harvard University Press

Graeber, Charles (2018), *The Breakthrough: Immunotherapy and the Race to Cure Cancer*, London: Scribe

Graeber, David (2016), *The Utopia of Rules: On Technology, Stupidity, and the Secret Joys of Bureaucracy*, New York: Melville House

Graeber, David (2018), *Bullshit Jobs: A Theory*, London: Allen Lane

Grant, Adam (2016), *Originals: How Non-Conformists Move the World*, New York: Viking

Grayling, A.C. (2021), *The Frontiers of Knowledge: What We Know About Science, History and The Mind – And How We Know It*, London: Viking

Greenspan, Alan, and Adrian Wooldridge (2018), *Capitalism in America: A History*, London: Allen Lane

Greif, Avner, Kiesling, Lynne, and Nye, John V.C. (eds) (2015), *Institutions, Innovation and Industrialization: Essays in Economic History and Development*, Princeton, NJ: Princeton University Press

Gross, Daniel P., and Sampat, Bhaven N. (2020), 'Inventing the Endless Frontier: The Effects of the World War II Research Effort on Post-War Innovation', Harvard Business School Strategy Unit Working Paper No. 20-126

Gruber, Jonathan, and Johnson, Simon (2019), *Jump-Starting America: How Breakthrough Science Can Revive Economic Growth and the American Dream*, New York: Public Affairs

Grush, Loren (2019), 'NASA's future Moon rocket will probably be delayed and over budget yet again: audit', The Verge, accessed 30 July 2019, available at https://www.theverge.com/2019/6/19/18691230/nasa-space-launch-system-orion-artemis-moon-human-exploration

Gusenbauer, M. (2019), 'Google Scholar to overshadow them all? Comparing the sizes of 12 academic search engines and bibliographic databases', *Scientometrics* 118, pp. 177–214

Guzey, Alexey (2019a), 'Reviving Patronage and Revolutionary Industrial Research', guzey.com, accessed 20 November 2019, available at https://guzey.com/patronage-and-research-labs/

Guzey, Alexey (2019b), 'Why We Likely Underappreciate the Pace of Technological Progress', guzey.com, accessed 12 November 2019, available at https://guzey.com/why-we-underappreciate-technological-progress/

Hafner, Danijar, Lillicrap, Timothy, Ba, Jimmy, and Norouzi, Mohammed (2019), 'Dream to Control: Learning Behaviors by Latent Imagination', arXiv, 1912.01603

Haldane, Andy (2015), 'Who Owns A Company?', Bank of England, accessed 2 April 2021, available at https://www.bankofengland.co.uk/-/media/boe/files/speech/2015/who-owns-a-company.pdf

Hanlon, Michael (2014), 'The golden quarter', Aeon, accessed 26 October 2018, available at https://aeon.co/essays/has-progress-in-science-and-technology-come-to-a-halt

Hanson, Robin (2016), *The Age of Em: Work, Love and Life when Robots Rule the Earth*, Oxford: Oxford University Press

Harari, Yuval Noah (2014), *Sapiens: A Brief History of Humankind*, London: Harvill Secker

Harari, Yuval Noah (2016), *Homo Deus: A Brief History of Tomorrow*, London: Harvill Secker

Harari, Yuval Noah (2019), *21 Lessons for the 21st Century*, London: Vintage

Hargadon, Andrew (2003), *How Breakthroughs Happen: The Surprising Truth About How Companies Innovate*, Boston, MA: Harvard Business School Publishing

Hassabis, Demis (2019), 'AI's potential', *The Economist*, accessed 5 February 2020, available at https://worldin.economist.com/article/17385/edition2020demis-hassabis-predicts-ai-will-supercharge-science

Hawksworth, John, Clarry, Rob, and Audino, Hannah (2017), 'The Long View: How will the global economic order change by 2050?', PwC, accessed 10 January 2021, available at https://www.pwc.com/gx/en/world-2050/assets/pwc-the-world-in-2050-full-report-feb-2017.pdf

Henderson, Caspar (2017), *A New Map of Wonders: A Journey in Search of Modern Marvels*, London: Granta

Henrich, Joseph (2009), 'The Evolution of Innovation-Enhancing Institutions'. In *Innovation in Cultural Systems: Contributions in Evolution Anthropology*, ed. Stephen J. Shennan and Michael J. O'Brien, Cambridge, MA: MIT Press

Henrich, Joseph (2016), *The Secret of Our Success: How Culture Is Driving Human Evolution, Domesticating Our Species, And Making Us Smarter*, Princeton, NJ: Princeton University Press

Henrich, Joseph (2020), *The Weirdest People in the World: How the West Became Psychologically Peculiar and Particularly Prosperous*, London: Allen Lane

Hesse, Herman (2000), *The Glass Bead Game*, London: Vintage

Higgs, John (2019), *The Future Starts Here: Adventures in the Twenty-First Century*, London: Weidenfeld & Nicolson

Hockfield, Susan (2019), *The Age of Living Machines: How Biology Will Build the Next Technology Revolution*, New York: W.W. Norton

Homer-Dixon, Thomas (2001), *The Ingenuity Gap: How Can We Solve the Problems of the Future?*, London: Vintage

Hooper, Rowan (2021), *How To Spend A Trillion Dollars: Saving the World and Solving the Biggest Mysteries in Science*, London: Profile Books

Hopenhayn, Hugo A., Neira, Julian, and Singhania, Rish (2018), 'From Population Growth to Firm Demographics: Implications for Concentration, Entrepreneurship and the Labor Share', NBER Working Paper No. w25382

Horgan, John (2015), *The End of Science: Facing the Limits of Knowledge in the Twilight of the Scientific Age*, New York: Basic Books

Hossenfelder, Sabine (2018), 'The Present Phase of Stagnation in the Foundations of Physics Is Not Normal', Nautil.us, accessed 22 December 2020, available at http://nautil.us/blog/the-present-phase-of-stagnation-in-the-foundations-of-physics-is-not-normal

Howell, Sabrina T., Lerner, Josh, Nanda, Ramana, and Townsend, Richard (2020), 'Financial Distancing: How Venture Capital

Follows the Economy Down and Curtails Innovation', Harvard Business School Entrepreneurial Management Working Paper No. 20-115

Howes, Anton (2017), 'The Spread of Improvement: Why Innovation Accelerated in Britain 1547–1851', accessed 19 December 2019, available at https://antonhowes.weebly.com/uploads/2/1/0/8/21082490/spread_of_improvement_working_paper.pdf

Howes, Anton (2019), 'The Crucial Century', antonhowes.com, accessed 16 March 2020, available at https://www.antonhowes.com/blog/the-crucial-century

Howes, Anton (2020), 'The Paradox of Progress', *The Age of Invention*, accessed 7 January 2021, available at https://antonhowes.substack.com/p/age-of-invention-the-paradox-of-progress

Howes, Anton (2021), 'Upstream, Downstream', *The Age of Invention*, accessed January 23, 2021, available at https://antonhowes.substack.com/p/age-of-invention-upstream-downstream

Huebner, Jonathan (2005), 'A possible declining trend for worldwide innovation', *Technological Forecasting and Social Change*, Vol. 72 No. 8, pp. 980–6

Hunt, Jennifer, and Gauthier-Loiselle, Marjolaine (2010), 'How Much Does Immigration Boost Innovation?', *American Economic Journal: Macroeconomics*, Vol. 2 No. 2, pp. 31–56.

Hvide, Hans K., and Jones, Benjamin F. (2018), 'University Innovation and the Professor's Privilege', *American Economic Review*, Vol. 108 No. 7, pp. 1860–98

IEA (2019), 'World Energy Investment 2019', IEA, Paris, accessed 19 October 2019, available at https://www.iea.org/reports/world-energy-investment-2019

Illies, Florian (2013), *1913: The Year Before the Storm*, London: The Clerkenwell Press

Ingham, Tim (2019), 'Nearly 40,000 Tracks Are Being Added To Spotify Every Single Day', *Music Business Worldwide*, accessed 15 September 2019, available at https://www.musicbusinessworldwide.com/nearly-40000-tracks-are-now-being-added-to-spotify-every-single-day/

Ioannidis, John (2005), 'Why Most Published Research Findings Are False', *Public Library of Science Medicine*, Vol. 2 No. 8: e124

Ioannidis, John (2018), 'Meta-research: Why research on research matters', *Public Library of Science Biology*, Vol. 16 No. 3: e2005468

Israel, Jonathan I. (2006), *Enlightenment Contested: Philosophy, Modernity and the Emancipation of Man 1670–1752*, Oxford: Oxford University Press

Jacobs, Margaret C. (2014), *The First Knowledge Economy: Human Capital and the European Economy, 1750–1850*, Cambridge: Cambridge University Press

Jaffe, Adam B., and Jones, Benjamin F. (eds) (2015), *The Changing Frontier: Rethinking Science and Innovation Policy*, Chicago: University of Chicago Press

Jaffe, Adam B., and Lerner, Josh (2004), *Innovation and Its Discontents: How Our Broken Patent System Is Endangering Innovation and Progress, And What To Do About It*, Princeton, NJ: Princeton University Press

Janik, Allan, and Toulmin, Stephen (1973), *Wittgenstein's Vienna*, New York: Simon and Schuster

Jockers, Matthew L. (2013), *Macroanalysis: Digital Methods and Literary History*, Champaign, IL: The University of Illinois Press

Johnson, Steven (2011), *Where Good Ideas Come From: The Seven Patterns of Innovation*, London: Penguin

Jones, Benjamin F. (2009), 'The Burden of Knowledge and the "Death of the Renaissance Man": Is Innovation Getting Harder?', *The Review of Economic Studies*, Vol. 76 No. 1, pp. 283–317

Jones, Benjamin F. (2010), 'Age and Great Invention', *The Review of Economics and Statistics*, Vol. 92 No. 1, pp. 1–14

Jones, Benjamin F., and Weinberg, Bruce A. (2011), 'Age dynamics in scientific creativity', *Proceedings of the National Academy of Sciences*, Vol. 108 No. 47

Jones, Benjamin F., Reedy, E.J., and Weinberg, Bruce A. (2014), 'Age and Scientific Genius', in *Handbook of Genius*, ed. Simonton, Dean, Sussex: Wiley

Jones, Charles I. (1995), 'R & D-Based Models of Economic Growth', *Journal of Political Economy*, Vol. 103 No. 4, pp. 759–84

Jones, Charles I. (2002), 'Sources of U.S. Economic Growth in a World of Ideas', *American Economic Review*, Vol. 92 No. 1, pp. 220–39

Jones, Charles I. (2019), 'Paul Romer: Ideas, Nonrivalry, and Endogenous Growth', *The Scandinavian Journal of Economics*, Vol. 121 No. 3, pp. 859–83

Jones, Charles I., and Vollrath, Dietrich (2013), *Introduction to Economic Growth: Third Edition*, New York: W.W. Norton

Joyner, Michael J., Paneth, Nigel, and Ioannidis, John P.A. (2016), 'What Happens When Underperforming Big Ideas in Research Become Entrenched?', *JAMA*, Vol. 316 No. 13, pp. 1355–6

Kahn, Herman, and Wiener, Anthony J. (1967), *The Year 2000: A Framework for Speculation on the Next Thirty-Three Years*, London: Macmillan

Kaku, Michio (2012), *Physics of the Future: The Inventions That Will Transform Our Lives*, London: Penguin

Kaku, Michio (2015), *The Future of the Mind: The Scientific Quest to Understand, Enhance and Empower the Mind*, London: Penguin

Kaku, Michio (2018), *The Future of Humanity: Terraforming Mars, Interstellar Travel, Immortality and Our Destiny Beyond Earth*, London: Allen Lane

Kardashev, Nikolai S. (1964), 'Transmission of Information by Extraterrestrial Civilizations', *Soviet Astronomy*, Vol. 8 No. 2, pp. 217–20

Kaufmann, Eric (2010), *Shall The Religious Inherit The Earth?: Demography and Politics in the Twenty-First Century*, London: Profile Books

Kay, John, and King, Mervyn (2020), *Radical Uncertainty*, London: The Bridge Street Press 2020

Kean, Sam (2019), 'A storied Russian lab is trying to push the periodic table past its limits – and uncover exotic new elements', *Science*, accessed 5 October 2019, available at https://www.sciencemag.org/news/2019/01/storied-russian-lab-trying-push-periodic-table-past-its-limits-and-uncover-exotic-new

Kelly, Bryan T., Papanikolaou, Dimitris, Seru, Amit, and Taddy, Matt (2020), 'Measuring Technological Innovation Over the Long Run', Yale ICF Working Paper No. 2018–19, available at SSRN: https://ssrn.com/abstract=3279254 or http://dx.doi.org/10.2139/ssrn.3279254

Kelly, Kevin (2017), *The Inevitable: Understanding The 12 Technological Forces That Will Shape Our Future*, New York: Penguin

Khanna, Parag (2019), *The Future Is Asian: Global Order in the 21st-Century Century*, London: Weidenfeld & Nicolson

Kim, K.H. (2017), 'The Creativity Crisis: It's Getting Worse', *Idea to Value*, accessed 26 August 2019, available at https://www.ideatovalue.com/crea/khkim/2017/04/creativity-crisis-getting-worse/

Kings Fund (2020), 'The NHS budget and how it has changed', accessed 9 January 2021, available at https://www.kingsfund.org.uk/projects/nhs-in-a-nutshell/nhs-budget

Kirsner, Scott (2018), 'The Biggest Obstacles to Innovation in Large Companies', *Harvard Business Review*, accessed 20 November 2019, available at https://hbr.org/2018/07/the-biggest-obstacles-to-innovation-in-large-companies

Klinger, Joel, Mateos-Garcia, Juan C., and Stathoulopoulos, Konstantinos (2020), 'A Narrowing of AI Research?', available at SSRN: https://ssrn.com/abstract=3698698

Knott, Anne Marie (2017), *How Innovation Really Works: Using the Trillion-Dollar R&D Fix to Drive Growth*, New York: McGraw-Hill

Koestler, Arthur (1964), *The Sleepwalkers: A History of Man's Changing Vision of the Universe*, London: Penguin

Koestler, Arthur (1970), *The Act of Creation*, London: Pan Piper

Kounios, John, and Mark Beeman (2015), *The Eureka Factor: Creative Insights and the Brain*, London: Windmill

Kremer, Michael (1993), 'Population Growth and Technological Change: One Million B.C. to 1990', *The Quarterly Journal of Economics*, Vol. 108 No. 3, pp. 681–716

Kuhn, Thomas S. (2012), *The Structure of Scientific Revolutions* (fourth edition), Chicago: University of Chicago Press

Lafond, François, and Kim, Daniel (2017), 'Long-Run Dynamics of the U.S. Patent Classification System', available at SSRN: https://ssrn.com/abstract=2924387

Laincz, C.A. and Peretto, P.F. (2006), 'Scale effects in endogenous growth theory: an error of aggregation not specification', *Journal of Economic Growth*, 11, pp. 263–88

Lakatos, Imre (1980), *The Methodology of Scientific Research Programmes: Philosophical Papers Volume 1*, Cambridge: Cambridge University Press

Landhuis, E. (2016), 'Scientific literature: Information overload', *Nature* 535, pp. 457–8

Le Fanu, James (2010), 'Science's dead end', *Prospect*, accessed 26 August 2019, available at https://www.prospectmagazine.co.uk/magazine/sciences-dead-end

Le Fanu, James (2011), *The Rise and Fall of Modern Medicine*, London: Abacus

Le Fanu, James (2018), *Too Many Pills: How Too Much Medicine is Endangering our Health and What We Can Do about It*, London: Little, Brown

Lehrer, Jonah (2012), *Imagine: How Creativity Works*, New York: Houghton Mifflin Harcourt

Lem, Stanisław (1974), *The Futurological Congress*, London: Penguin Modern Classics

Lenton, Tim et al. (2019), 'Climate tipping points – too risky to bet against', *Nature*, accessed 5 January 2020, available at https://www.nature.com/articles/d41586-019-03595-0

Lingfei Wu, Dashun, Wang, and Evans, James A. (2019), 'Large teams develop and small teams disrupt science and technology', *Nature*, 566, pp. 378–82

Lloyd, Seth (2001), 'Computational capacity of the universe', *CERN*, accessed 12 January 2021, available at http://cds.cern.ch/record/524220/files/0110141.pdf

Loeb, Avi (2021), *Extraterrestrial: The First Sign of Intelligent Life Beyond Earth*, London: John Murray

Lovejoy, Arthur O. (1976), *The Great Chain of Being: A Study of the History of An Idea*, Cambridge, MA: Harvard University Press

Lovelock, James (2014), *A Rough Ride to the Future*, London: Allen Lane

Lovelock, James (2019), *Novacene: The Coming Age of Hyperintelligence*, London: Allen Lane

Lucking, Brian, Bloom, Nicholas, and Van Reenen, John (2018), 'Have R&D Spillovers Changed?', NBER Working Paper No. w24622

Lukianoff, Greg, and Haidt, Jonathan (2018), *The Coddling of the American Mind: How Good Intentions and Bad Ideas are Setting Up a Generation for Failure*, London: Allen Lane

Maçães, Bruno (2021), 'After Covid, get ready for the Great Acceleration', *Spectator*, accessed 15 March 2021, available at https://www.spectator.co.uk/article/after-covid-get-ready-for-the-great-acceleration

Maddox, John (1999), *What Remains to Be Discovered: Mapping the Secrets of the Universe, the Origins of Life, and the Future of the Human Race*, New York: Touchstone

Madrigal, Alexis C. (2020), 'Silicon Valley Abandons the Culture That Made It the Envy of the World', *The Atlantic*, accessed 21 January 2020, available at https://www.theatlantic.com/technology/archive/2020/01/why-silicon-valley-and-big-tech-dont-innovate-anymore/604969/

Mahbubani, Kishore (2018), *Has The West Lost It?: A Provocation*, London: Allen Lane

Mahon, Basil (2004), *The Man Who Changed Everything: The Life of James Clerk Maxwell*, Chichester: Wiley

Mallapaty, Smriti (2018), 'Paper authorship goes hyper', *Nature Index*, accessed 19 October 2018, available at https://www.natureindex.com/news-blog/paper-authorship-goes-hyper

Malone, Thomas W. (2018), *Superminds: How Hyperconnectivity is Changing the Way We Solve Problems*, London: Oneworld

Marcus, Jon (2018), 'With enrollment sliding, liberal arts colleges struggle to make a case for themselves', *The Hechinger Report*, accessed 9 January 2021, available at https://hechingerreport.org/with-enrollment-sliding-liberal-arts-colleges-struggle-to-make-a-case-for-themselves/

Martin, Richard (2015), 'Weighing the Cost of Big Science',

MIT Technology Review, accessed 6 March 2019, available at https://www.technologyreview.com/2015/09/22/166155/weighing-the-cost-of-big-science/

Martinis, John, and Boixo, Sergio (2019), 'Quantum Supremacy Using a Programmable Superconducting Processor', Google Blog, accessed 11 April 2020, available at https://ai.googleblog.com/2019/10/quantum-supremacy-using-programmable.html?m=1

Mazzucato, Mariana (2013), *The Entrepreneurial State: Debunking Public vs. Private Sector Myths*, London: Anthem Press

Mazzucato, Mariana (2018a), 'Mission-oriented innovation policies: challenges and opportunities', *Industrial and Corporate Change*, Vol. 27 No. 5, pp. 803–15

Mazzucato, Mariana (2018b), *The Value of Everything: Making and Taking in the Global Economy*, London: Allen Lane

Mazzucato, Mariana (2021), *Mission Economy: A Moonshot Guide to Changing Capitalism*, London: Allen Lane

McAfee, Andrew, and Brynjolfsson, Erik (2017), *Machine Platform Crowd: Harnessing Our Digital Future*, New York: W.W. Norton

McCloskey, Deirdre Nansen (2017), *Bourgeois Equality: How Ideas, Not Capital or Institutions, Enriched the World*, Chicago: University of Chicago Press

McCormick, Packy (2020), 'Conjuring Scenius', perrell.com, accessed 18 June 2020, available at https://perell.com/fellowship/conjuring-scenius/

McCullough, David (2015), *The Wright Brothers: The Dramatic Story behind the Legend*, London: Simon & Schuster

McDonald, Ian (2005), *River of Gods*, London: Pocket Books

McKinsey Global Institute (2020), 'The Bio Revolution: Innovations transforming economies, societies, and our lives', *McKinsey Global Institute*, accessed 27 May 2020, available at https://www.mckinsey.com/industries/pharmaceuticals-and-medical-products/our-insights/the-bio-revolution-innovations-transforming-economies-societies-and-our-lives#

McMahon, Jeff (2020), 'Is Fusion Really Close To Reality? Yes, Thanks To Machine Learning', *Forbes*, accessed 23

September 2020, available at https://www.forbes.com/sites/jeffmcmahon/2020/04/27/is-fusion-really-closer-to-reality-yes-thanks-to-machine-learning/?sh=4653592652b6

Meng, Fankang, and Ellis, Tom (2020), 'The second decade of synthetic biology: 2010–2020', *Nature Communications*, Vol. 11 No. 5174

Merton, R. (1961), 'Singletons and Multiples in Scientific Discovery: A Chapter in the Sociology of Science', *Proceedings of the American Philosophical Society*, Vol. 105 No. 5, pp. 470–86, accessed 8 September 2020, available from http://www.jstor.org/stable/985546

Merton, Robert K. (1996), *On Social Structure and Science*, Chicago: University of Chicago Press

Merton, Robert K., and Barber, Elinor (2003), *The Travels and Adventures of Serendipity*, Princeton, NJ: Princeton University Press

Mesoudi, Alex (2011), 'Variable Cultural Acquisition Costs Constrain Cumulative Cultural Evolution', *Public Library of Science ONE*, Vol. 6 No. 3, e18239

Miller, Arthur I. (2019), *The Artist in the Machine: The World of AI-Powered Creativity*, Boston, MA: MIT Press

Mlodinow, Leonard (2019), *Elastic: Flexible Thinking in a Constantly Changing World*, London: Penguin

Mokyr, Joel (1990), *The Lever of Riches: Technological Creativity and Economic Progress*, New York: Oxford University Press

Mokyr, Joel (1994), 'Cardwell's Law and the political economy of technological progress', *Research Policy*, Vol. 23 No. 5, pp. 561–74

Mokyr, Joel (1999), 'The Second Industrial Revolution, 1870–1914', in *Storia dell'economia Mondiale*, Rome: Laterza

Mokyr, Joel (2002), *The Gifts of Athena: Historical Origins of the Knowledge Economy*, Princeton, NJ: Princeton University Press

Mokyr, Joel (2011), *The Enlightened Economy: Britain and the Industrial Revolution 1700–1850*, London: Penguin

Mokyr, Joel (2014), 'Big Ideas: Riding the Technology Dragon', *The Milken Institute Review*, Second Quarter 2014

Mokyr, Joel (2017), *A Culture of Growth: The Origins of the Modern Economy*, Princeton, NJ: Princeton University Press

Mokyr, Joel (2018), 'The past and the future of innovation: Some lessons from economic history', *Explorations in Economic History*, Vol. 69, pp. 13–26,

Moore, Gillian (2019), *The Rite of Spring: The Music of Modernity*, London: Head of Zeus

Moretti, Franco (2013), *Distant Reading*, London: Verso

Moretti, Enrico (2019), 'The Effect of High-Tech Clusters on the Productivity of Top Inventors', NBER Working Paper 26270

Morieux, Yves (2017), 'Technology is improving, productivity isn't. Why?', *Brunswick Review*, accessed 4 January 2019, available at https://www.brunswickgroup.com/yves-morieux-i6394/

Morris, Ian (2010), *Why The West Rules – For Now: The patterns of history and what they reveal about the future*, London: Profile Books

Morris, Ian (2014), *War, What Is It Good For?: The Role of Conflict in Civilisation, From Primates to Robots*, London: Profile Books

Morton, Oliver (2019), 'The engineering of living organisms could soon start changing everything', *The Economist*, accessed 3 January 2020, available at https://www.economist.com/technology-quarterly/2019/04/04/the-engineering-of-living-organisms-could-soon-start-changing-everything

Mulgan, Geoff (2018), *Big Mind: How Collective Intelligence Can Change The World*, Princeton, NJ: Princeton University Press

Muller, Jerry Z. (2018), *The Tyranny of Metrics*, Princeton, NJ: Princeton University Press

Muthukrishna, Michael, and Henrich, Joseph (2016), 'Innovation in the collective brain', *Philosophical Transactions of the Royal Society B*, Vol. 371 No. 1690, 20150192

National Science Board (2020), 'The State of U.S. Science and Engineering 2020', National Science Foundation, accessed 12 January 2021, available at https://ncses.nsf.gov/pubs/nsb20201/preface

Naudé, Wim (2019), 'The surprising decline of entrepreneurship and innovation in the West', *The Conversation*, accessed 11 April 2020, available at https://theconversation.com/the-surprising-decline-of-entrepreneurship-and-innovation-in-the-west-124552

Nichols, Tom (2017), *The Death of Expertise: The Campaign Against Established Knowledge and Why it Matters*, New York: Oxford University Press

Nielsen, Michael (2012), *Reinventing Discovery: The New Era of Networked Science*, Princeton, NJ: Princeton University Press

Nisbett, Richard E. (2003), *The Geography of Thought: How Asians and Westerners Think Differently – And Why*, London: Nicholas Brealey

Norberg, Johan (2020), *Open: The Story of Human Progress*, London: Atlantic Books

O'Connor, Cailin, and Weatherall, James Owen (2018), 'Scientific polarization', *European Journal for Philosophy of Science*, Vol. 8 No. 3, pp. 855–75

O'Mahony, Seamus (2019), *Can Medicine Be Cured? The Corruption of a Profession*, London: Head of Zeus

Odlyzko, Andrew (1995), 'The decline of unfettered research', University of Minnesota, Twin Cities, accessed 11 October 2020, available at http://www.dtc.umn.edu/~odlyzko/doc/decline.txt

OECD (2020), 'Gross domestic spending on R&D', OECD, accessed 17 August 2020, available at https://data.oecd.org/rd/gross-domestic-spending-on-r-d.htm

Office of National Statistics (2015), 'How has life expectancy changed over time?', Office of National Statistics, accessed 14 April 2019, available at https://www.ons.gov.uk/peoplepopulationandcommunity/birthsdeathsandmarriages/lifeexpectancies/articles/howhaslifeexpectancychangedovertime/2015-09-09

Okoshi, Yuki (2019), 'China's research papers lead the world in cutting-edge tech', Nikkei Asia, accessed 27 January 2021, available at https://asia.nikkei.com/Business/China-tech/China-s-research-papers-lead-the-world-in-cutting-edge-tech

Ovenden, Richard (2020), *Burning The Books: A History of Knowledge Under Attack*, London: John Murray

Packalen, Mikko, and Bhattacharya, Jay (2012), 'Words in Patents:

Research Inputs and the Value of Innovativeness in Invention', NBER Working Paper 18494

Packalen, Mikko, and Bhattacharya, Jay (2015), 'Cities and Ideas', NBER Working Paper No. w20921

Packalen, Mikko, and Bhattacharya, Jay (2019), 'Age and the Trying Out of New Ideas', *Journal of Human Capital*, Vol. 13 No. 2, pp. 341–73

Packalen, Mikko, and Bhattacharya, Jay (2020), 'NIH funding and the pursuit of edge science', *PNAS*, Vol. 117 No. 22

Parker, Ian (2020), 'Yuval Noah Harari's History of Everyone, Ever', *New Yorker*, accessed 13 March 2020, available at https://www.newyorker.com/magazine/2020/02/17/yuval-noah-harari-gives-the-really-big-picture

Perez, Carlota (2002), *Technological Revolutions and Financial Capital: The Dynamics of Bubbles and Golden Ages*, Cheltenham: Edward Elgar

Pilling, David (2018), *The Growth Delusion: The Wealth and Well-Being of Nations*, London: Bloomsbury

Pinker, Stephen (2019), *Enlightenment Now: The Case for Reason, Science, Humanism and Progress*, London: Penguin

Pinker, Stephen, Ridley, Matt, De Botton, Alain, and Gladwell, Malcolm (2016), *Do Humankind's Best Days Lie Ahead?*, Toronto: House of Anansi

Platt, John R. (1964), 'Strong Inference', *Science*, Vol. 146 No. 3642, pp. 347–53

Rees, Martin (2018), *On The Future: Prospects For Humanity*, Princeton, NJ: Princeton University Press

Reinhardt, Ben (2020), 'Notes on Are Ideas Getting Harder to Find?', benreinhardt.com, accessed 25 September 2020, available at https://benjaminreinhardt.com/notes-on-are-ideas-getting-harder-to-find?

Reinhardt, Ben (2021), *Shifting the impossible to the inevitable: A Private ARPA User Manual*, benreinhardt.com, accessed 12 April 2021, available at https://benjaminreinhardt.com/parpa

Reller, Tom (2016), 'Elsevier publishing – a look at the numbers, and more', Elsevier.com, accessed June 8, 2019, available at https://www.elsevier.com/connect/elsevier-publishing-a-look-at-the-numbers-and-more

Renwick, Chris (2017), *Bread For All: The Origins of the Welfare State*, London: Allen Lane

Reynolds, Matt (2020), 'DeepMind's AI is getting closer to its first big real-world application', *Wired*, accessed 5 February 2020, available at https://www.wired.co.uk/article/deepmind-protein-folding-alphafold

Ricón, José Luis (2015), 'Is there R&D spending myopia?', *Nintil*, accessed 6 January 2021, available at https://nintil.com/is-there-rd-spending-myopia/

Ricón, José Luis (2019), 'On Bloom's two sigma problem: A systematic review of the effectiveness of mastery learning, tutoring, and direct instruction', *Nintil*, accessed 20 July 2020, available at https://nintil.com/bloom-sigma/

Ricón, José Luis (2020a), 'Fund people, not projects I: The HHMI and the NIH Director's Pioneer Award', *Nintil*, accessed 24 January 2021, available at https://nintil.com/hhmi-and-nih/

Ricón, José Luis (2020b), 'Was Planck right? The effects of aging on the productivity of scientists', *Nintil*, accessed 14 January 2021, available at https://nintil.com/age-and-science/

Ridley, Matt (2011), *The Rational Optimist: How Prosperity Evolves*, London: Fourth Estate

Ridley, Matt (2016), *The Evolution of Everything: How Small Changes Transform Our World*, London: Fourth Estate

Ridley, Matt, (2020), *How Innovation Works*, London: Fourth Estate

Rifkin, Jeremy (2014), *The Zero Marginal Cost Society: The Internet of Things, The Collaborative Commons, and The Eclipse of Capitalism*, New York: Palgrave

Ringel, Michael S., Scannell, Jack, Baedeker, Mathias, and Schulze, Ulrik (2020), 'Breaking Eroom's Law', *Nature Reviews Drug Discovery*, Vol. 19, pp. 833–4

Ritchie, Stuart (2020), *Science Fictions: Exposing Fraud, Bias, Negligence and Hype in Science*, London: Bodley Head

Robertson, Ritchie (2020), *The Enlightenment: The Pursuit of Happiness, 1680–1790*, London: Allen Lane

Robinson, Andrew (2006), *The Last Man Who Knew Everything: Thomas Young, the Anonymous Polymath Who Proved Newton Wrong, Explained How We See, Cured the Sick and Deciphered the Rosetta Stone*, London: Oneworld

Rockey, Sally (2015), 'More Data on Age and the Workforce', *NIH Extramural News*, accessed 25 August 2019, available at https://nexus.od.nih.gov/all/2015/03/25/age-of-investigator/

Rodriques, Samuel G., and Marblestone, Adam H. (2020), 'Focused Research Organizations to Accelerate Science, Technology, and Medicine', *Day One Project*, accessed 15 November 2020, available at https://www.dayoneproject.org/post/focused-research-organizations-to-accelerate-science-technology-and-medicine

Rogers Hollingsworth, J. (2007), 'High Cognitive Complexity and the Making of Major Scientific Discoveries', in *Knowledge, Communication and Creativity*, Los Angeles: SAGE

Romer, Paul M. (1990), 'Endogenous Technological Change', *The Journal of Political Economy*, Vol. 98 No. 5, Part 2: The Problem of Development: A Conference of the Institute for the Study of Free Enterprise Systems

Roser, Max (2020), 'Why is life expectancy in the US lower than in other rich countries?', Our World in Data, accessed 10 November 2020, available at https://ourworldindata.org/us-life-expectancy-low

Rosling, Hans, with Rosling, Ola, and Rosling, Anna Rönnlund (2018), *Factfulness: Ten Reasons We're Wrong About The World – And Why Things Are Better Than You Think*, London: Sceptre

Ross, Alex (2017), *The Industries of the Future*, London: Simon & Schuster

Rotman, David (2019), 'AI is reinventing the way we invent', *MIT Technology Review*, accessed 26 March 2019, available at https://www.technologyreview.com/2019/02/15/137023/ai-is-reinventing-the-way-we-invent/

Roy, Avik (2012), *Stifling New Cures: The True Cost of Lengthy Clinical Drug Trials*, New York: Manhattan Institute

Sand, Shlomo (2018), *The End of the French Intellectual: From Zola to Houellebecq*, London: Verso

Sandberg, Anders, Drexler, Eric, and Ord, Toby (2018), 'Dissolving the Fermi Paradox', arXiv, 1806.02404

Sarewitz, Dan (2016), 'Saving Science', *The New Atlantis*, accessed 24 December 2019, available at https://www.thenewatlantis.com/publications/saving-science

Sargent, John F. (2020), 'Global Research and Development Expenditures', *Congressional Research Service*, accessed 31 May 2020, available at https://fas.org/sgp/crs/misc/R44283.pdf

Sautoy, Marcus du (2017), *What We Cannot Know: From Consciousness to the Cosmos, the Cutting Edge of Science Explained*, London: Fourth Estate

Sautoy, Marcus du (2019), *The Creativity Code: How AI Is Learning to Write, Paint and Think*, London: Fourth Estate

Scannell, J., Blanckley, A., Boldon, H. et al. (2012), 'Diagnosing the decline in pharmaceutical R&D efficiency', *Nature Reviews Drug Discovery*, Vol. 11, pp. 191–200

Scharf, Caleb (2015), *The Copernicus Complex: The Quest for Our Cosmic (In)Significance*, London: Penguin

Scheu, René (2019), 'PayPal founder and philosopher Peter Thiel: "The heads in Silicon Valley have aligned themselves"', *Neue Zürcher Zeitung*, accessed 9 April 2019, available at https://www.nzz.ch/feuilleton/peter-thiel-donald-trump-handelt-fuer-mich-zu-wenig-disruptiv-ld.1471818?reduced=true

Schwab, Klaus (2017), *The Fourth Industrial Revolution*, London: Portfolio Penguin

Schwartz, Peter (1991), *The Art of the Long View: Planning for the Future in an Uncertain World*, Chichester: John Wiley

Senior, Andrew, Jumper, John, Hassabis, Demis, and Kohli, Pushmeet (2020), 'AlphaFold: Using AI for scientific discovery', DeepMind, accessed 5 February 2020, available at https://deepmind.com/blog/article/AlphaFold-Using-AI-for-scientific-discovery

Shambaugh, Jay, Nunn, Ryan, Breitwieser, Audrey, and Liu, Patrick (2018), *The State of Competition and Dynamism: Facts about Concentration, Start-Ups, and Related Policies*, Washington DC: The Hamilton Project

Shaxson, Nicholas (2018), *The Finance Curse: How Global Finance Is Making Us All Poorer*, London: The Bodley Head

Sheldrake, Rupert (2013), *The Science Delusion: Freeing the Spirit of Enquiry*, London: Coronet

Shi, Feng, and Evans, James (2019), 'Science and Technology Advance through Surprise', arXiv, 1910.09370

Simonite, Tom (2014), 'Technology Stalled in 1970', *MIT Technology Review*, accessed 14 July 2019, available at https://www.technologyreview.com/2014/09/18/171322/technology-stalled-in-1970/

Singh, Jasjit, and Fleming, Lee (2010), 'Lone Inventors as Sources of Breakthroughs: 'Myth or Reality?', *Management Science*, Vol. 56 No. 1

Skinner, Michael (2016), 'Unified theory of evolution', Aeon, accessed November 2, 2019, available at https://aeon.co/essays/on-epigenetics-we-need-both-darwin-s-and-lamarck-s-theories

Sloman, Steven, and Fernbach, Philip (2017), *The Knowledge Illusion: Why We Never Think Alone*, London: Macmillan

Smil, Vaclav (2005), *Creating the Twentieth Century: Technical Innovations of 1867–1916 and Their Lasting Impact*, New York: Oxford University Press

Smil, Vaclav (2017), *Energy and Civilization: A History*, Cambridge, MA: MIT Press

Smil, Vaclav (2020), *Numbers Don't Lie: 71 Things You Need to Know About the World*, London: Viking

Smith, Kendall A. (2012), 'Louis Pasteur: The Father of Immunology', *Frontiers of Immunology*, Vol. 3 No. 86

Smith, Noah (2021), 'Interview: Patrick Collison, co-founder and CEO of Stripe', *Noahpinion*, accessed 29 March 2021, available at https://noahpinion.substack.com/p/interview-patrick-collison-co-founder

Smolin, Lee (2008), *The Trouble with Physics: The Rise of String Theory, the Fall of a Science and What Comes Next*, London: Penguin

Smolin, Lee (2019), *Einstein's Unfinished Revolution: The Search for What Lies Beyond the Quantum*, London: Allen Lane

Solms, Mark (2021), *The Hidden Spring: A Journey to the Source of Consciousness*, London: Profile Books

Solow, Robert M. (1956), 'A Contribution to the Theory of

Economic Growth', *The Quarterly Journal of Economics*, Vol. 70 No. 1, pp. 65–94

Southwood, Ben (2020), 'The rise and fall of the industrial R&D lab', *Works in Progress*, accessed 11 November 2019, available at https://worksinprogress.co/issue/the-rise-and-fall-of-the-american-rd-lab/

Spengler, Oswald (1991), *The Decline of the West*, Oxford: Oxford University Press

Stern, Nicholas (2015), *Why Are We Waiting?: The Logic, Urgency and Promise of Tackling Climate Change*, Cambridge, MA: MIT Press

Stokes, Jonathan M. (2020), 'A Deep Learning Approach to Antibiotic Discovery', *Cell*, Vol. 180 No. 4, pp. 688–702.e13

Storrs Hall, J. (2018), *Where Is My Flying Car? A Memoir of Future Past*, independently published

Sverdlov, Eugene (2018), 'Incremental Science: Papers and Grants, Yes; Discoveries, No', *Molecular Genetics Microbiology and Virology*, Vol. 33 No. 4, pp. 207–16

Syed, Matthew (2019), *Rebel Ideas: The Power of Diverse Thinking*, London: John Murray

Szell, Michael, Ma, Yifang, and Sinatra, Roberta (2018), 'A Nobel opportunity for interdisciplinarity', *Nature Physics*, 14, pp. 1075–8

Tainter, Joseph A. (1988), *The Collapse of Complex Societies*, Cambridge: Cambridge University Press

Teare, Gene (2020), 'Diversity Report: 20 Percent Of Newly Funded Startups In 2019 Have A Female Founder', *Crunchbase*, accessed 12 January 2021, available at https://news.crunchbase.com/news/eoy-2019-diversity-report-20-percent-of-newly-funded-startups-in-2019-have-a-female-founder/

Tegmark, Max (2017), *Life 3.0: Being human in an age of Artificial Intelligence*, London: Allen Lane

Tetlock, Philip, E., and Gardner, Dan (2015), *Superforecasting: The Art and Science of Prediction*, New York: Crown

The Economist (2016), 'How Vienna produced ideas that shaped the West', *The Economist*, accessed 8 December 2019, available at https://www.economist.com/christmas-specials/2016/12/24/how-vienna-produced-ideas-that-shaped-the-west

Therrien, Alex (2018), 'Life expectancy progress in UK "stops for first time"', BBC, accessed 14 April 2019, available at https://www.bbc.co.uk/news/health-45638646

Thiel, Peter (2014), *Zero to One: Notes on Startups, Or How to Build the Future*, London: Virgin Books

Thompson, Michael J., and Smulewicz-Zucker, Gregory R. (eds) (2018), *Anti-Science and the Assault on Democracy*, New York: Prometheus Books

Thompson, Neil C., Greenewald, Kristjan, Lee, Keeheon, and Manso, Gabriel F. (2020), 'The Computational Limits of Deep Learning', arXiv, 2007.05558

Tiner, John Hudson (1990), *Louis Pasteur: Founder of Modern Medicine*, Fenton, MI: Mott Media

Tollefson, Jeff (2020), 'How Trump damaged science – and why it could take decades to recover', *Nature*, accessed 22 December 2020, available at https://www.nature.com/articles/d41586-020-02800-9

Tshitoyan, Vahe, Dagdelen, John, Weston, Leigh, et al. (2019), 'Unsupervised word embeddings capture latent knowledge from materials science literature', *Nature*, 571, pp. 95–8

Tuchman, Barbara (1985), *The March of Folly: From Troy To Vietnam*, New York: Random House

Urquiola, Miguel (2020), *A College on a Hill: Why America Leads the World in University Research*, Cambridge, MA: Harvard University Press

Uzzi, Brian, Mukherjee, Satyam, Stringer, Michael, and Jones, Benjamin F. (2013), 'Atypical Combinations and Scientific Impact', *Science*, 342, p. 468

Verhoeven, Dennis (2020), 'Potluck or Chef de Cuisine? Knowledge Diversity in Teams and Breakthrough Invention', available at SSRN: https://ssrn.com/abstract=2629602

Vijg, Jan (2011), *The American Technological Challenge: Stagnation and Decline in the 21st Century*, New York: Algora Publishing

Vince, Gaia (2019), *Transcendence: How Humans Evolved through Fire, Language, Beauty and Time*, London: Allen Lane

Vollrath, Dietrich (2020a), *Fully Grown: Why a Stagnant Economy Is a Sign of Success*, Chicago: University of Chicago Press

Vollrath, Dietrich (2020b), 'When did productivity growth slow down?', *Growthecon*, accessed 29 March 2021, available at https://growthecon.com/blog/BLS-TFP/

Vollset, Stein Emil et al. (2020), 'Fertility, mortality, migration, and population scenarios for 195 countries and territories from 2017 to 2100', *The Lancet*, Vol. 396 No. 10258, pp. 1285–1306

Wallace-Wells, David (2019), *The Uninhabitable Earth: A Story of the Future*, London: Penguin

Warsh, David (2007), *Knowledge and the Wealth of Nations: A Story of Economic Discovery*, New York: W.W. Norton

Watney, Caleb (2020), 'Cracks in the Great Stagnation', *Agglomerations*, accessed 7 January 2021, available at https://www.agglomerations.tech/cracks-in-the-great-stagnation/

Watson, Peter (2001), *The Modern Mind: An intellectual history of the 20th century*, London: HarperCollins

Watson, Peter (2006), *Ideas: A History from Fire to Freud*, London: Phoenix

Watson, Peter (2017), *Convergence: The Idea at the Heart of Science*, London: Simon and Schuster

Weightman, Gavin (2015), *Eureka: How Invention Happens*, New Haven, CT: Yale University Press

Weiman, Carl (2017), *Improving How Universities Teach Science: Lessons from the Science Education Initiative*, Cambridge, MA: Harvard Business Review Press

Weitzman, Martin (1996), 'Hybridizing Growth Theory', *The American Economic Review*, Vol. 86 No. 2, pp. 207–12

Wellcome Trust (2020), 'What Researchers Think About the Culture They Work In', Wellcome.org, accessed 21 January 2020, available at https://wellcome.org/reports/what-researchers-think-about-research-culture?utm_source=twitterShare

White, Curtis (2003), *The Middle Mind: Why Americans Don't Think for Themselves*, San Francisco: HarperSanFrancisco

White, Curtis (2014), *The Science Delusion: Asking the Big Questions in a Culture of Easy Answers*, New York: Melville House

Whitehead, A.N. (1925), *Science and the Modern World*, London: Macmillan

Williams, Jeffrey J. (2018), 'The Rise of the Promotional Intellectual', *The Chronicle of Higher Education*, accessed 22 August 2018, available at https://www.chronicle.com/article/the-rise-of-the-promotional-intellectual/

Wilson, Edward O. (2017), *The Origins of Creativity*, London: Allen Lane

Winchester, Simon (2008), *The Man Who Loved China: The Fantastic Story of the Eccentric Scientist Who Unlocked the Mysteries of the Middle Kingdom*, New York: HarperCollins

Wolf, Martin (2019), 'On the Technological Slowdown', *Foreign Affairs*, accessed 14 July 2019, available at https://www.foreignaffairs.com/articles/2015-11-19/martin-wolf-innovation-slowdown

Wong, May (2017), 'Scholars say big ideas are getting harder to find', Phys.org, accessed 10 October 2018, available at https://phys.org/news/2017-09-scholars-big-ideas-harder.html

Wootton, David (2015), *The Invention of Science: A New History of the Scientific Revolution*, London: Allen Lane

Wright, Robert (2000), *Nonzero: History, Evolution and Human Cooperation*, New York: Pantheon Books

Wright, Ronald (2006), *A Short History of Progress*, Edinburgh: Canongate

Wu, L., Wang, D., and Evans, J.A. (2019), 'Large teams develop and small teams disrupt science and technology', *Nature* 566, pp. 378–82

Wuchty, Stefan, Jones, Benjamin F., and Uzzi, Brian (2007), 'The Increasing Dominance of Teams in Production of Knowledge', *Science*, Vol. 316 No. 5827, pp. 1036–9

Xinhua (2019), 'China to build scientific research station on Moon's south pole', Xinhua, accessed 18 January 2021, available at http://www.xinhuanet.com/english/2019-04/24/c_138004666.htm

Yueh, Linda (2018), *The Great Economists: How Their Ideas Can Help Us Today*, London: Penguin Viking

Index

Michael Bhaskar is a writer, researcher and digital publisher. He is co-founder of Canelo, a new publishing company, and was a consultant Writer in Residence at DeepMind, the world's leading AI research lab. He has written and talked extensively about the future of media, the creative industries and the economics of technology for newspapers, magazines and blogs. He has been featured in and written for the *Guardian*, *Financial Times* and *Daily Telegraph*, and on BBC 2, BBC Radio 4, NPR and Bloomberg TV among others. He has been a British Council Young Creative Entrepreneur and a Frankfurt Book Fair Fellow. He is also author of *Curation*, *The Content Machine* and the *Oxford Handbook of Publishing*.